IoT and WSN Applications for Modern Agricultural Advancements:

Emerging Research and Opportunities

Proshikshya Mukherjee
KIIT University (Deemed), India

Prasant Kumar Pattnaik
KIIT University (Deemed), India

Surya Narayan Panda
Chitkara University, India

T0321475

A volume in the Advances in
Wireless Technologies and
Telecommunication (AWTT) Book
Series

Published in the United States of America by
    IGI Global
    Engineering Science Reference (an imprint of IGI Global)
    701 E. Chocolate Avenue
    Hershey PA, USA 17033
    Tel: 717-533-8845
    Fax: 717-533-8661
    E-mail: cust@igi-global.com
    Web site: http://www.igi-global.com

Library of Congress Cataloging-in-Publication Data

Names: Mukherjee, Proshikshya, 1992- editor.
Title: IoT and WSN applications for modern agricultural advancements :
  emerging research and opportunities / Proshikshya Mukherjee, Prasant Kumar
  Pattnaik, and Surya Narayan Panda, editors.
Description: Hershey, PA : Engineering Science Reference, an imprint of IGI
  Global, [2019] | Includes bibliographical references.
Identifiers: LCCN 2018058998| ISBN 9781522590040 (hardcover) | ISBN
  9781522590057 (softcover) | ISBN 9781522590064 (ebook)
Subjects: LCSH: Agriculture--Data processing. | Agricultural innovations.
Classification: LCC S494.5.D3 I595 2019 | DDC 635.028/6--dc23 LC record available at https://
lccn.loc.gov/2018058998

This book is published in the IGI Global book series Advances in Wireless Technologies and Telecommunication (AWTT) (ISSN: 2327-3305; eISSN: 2327-3313)

British Cataloguing in Publication Data
A Cataloguing in Publication record for this book is available from the British Library.

All work contributed to this book is new, previously-unpublished material.
The views expressed in this book are those of the authors, but not necessarily of the publisher.

For electronic access to this publication, please contact: eresources@igi-global.com.

Advances in Wireless Technologies and Telecommunication (AWTT) Book Series

ISSN:2327-3305
EISSN:2327-3313

Editor-in-Chief: Xiaoge Xu, University of Nottingham Ningbo China, China

MISSION

The wireless computing industry is constantly evolving, redesigning the ways in which individuals share information. Wireless technology and telecommunication remain one of the most important technologies in business organizations. The utilization of these technologies has enhanced business efficiency by enabling dynamic resources in all aspects of society.

The **Advances in Wireless Technologies and Telecommunication Book Series** aims to provide researchers and academic communities with quality research on the concepts and developments in the wireless technology fields. Developers, engineers, students, research strategists, and IT managers will find this series useful to gain insight into next generation wireless technologies and telecommunication.

COVERAGE

- Wireless Sensor Networks
- Mobile Web Services
- Radio Communication
- Wireless Technologies
- Global Telecommunications
- Mobile Technology
- Wireless Broadband
- Virtual Network Operations
- Cellular Networks
- Grid Communications

IGI Global is currently accepting manuscripts for publication within this series. To submit a proposal for a volume in this series, please contact our Acquisition Editors at Acquisitions@igi-global.com or visit: http://www.igi-global.com/publish/.

Titles in this Series

For a list of additional titles in this series, please visit:
https://www.igi-global.com/book-series/advances-wireless-technologies-telecommunication/73684

For an entire list of titles in this series, please visit:
https://www.igi-global.com/book-series/advances-wireless-technologies-telecommunication/73684

701 East Chocolate Avenue, Hershey, PA 17033, USA
Tel: 717-533-8845 x100 • Fax: 717-533-8661
E-Mail: cust@igi-global.com • www.igi-global.com

To my Guide and Mentor, Dr. Prasant Kumar Pattnaik, KIIT Deemed to be University Bhubaneswar, India. The process of how you do research and made the research work simple and very much interesting if you learn the rules of research, I learned from him. My parents have been always my source of motivation, inspiration, and as a wonderful mentor, guide, advisor, philosopher, and supporter to be remembered all time. My friend, Srijib Banerjee, who has always supported me. – Proshikshya Mukherjee

To my wife, Bismita, and my daughter, Prasannakshi. – Prasant Kumar Pattnaik

To my daughter, Vaishali. – Surya Narayan Panda

Table of Contents

Chapter 5

Chapter 6

Chapter 7

Preface

An expert system is designed to solve complex problems which usually requires human expertise. It automates the decision making by representing mainly if-then-else rule rather than using conventional procedural methods. Inference engine and Knowledge base are mainly the two subsystems that an expert system can divide into. Knowledge base basically represents facts and rules whereas inference engine applies to the rules to the known facts. Expert system mainly implicates the increase in productivity via quick diagnoses and allowing the expert knowledge to be extracted when there is no ready access to human expert.

This book organized into seven chapters.

In the introduction, agriculture with technology and application of expert systems in various fields like wireless sensor network and IoT of engineering has been discussed.

Chapter 1 surveys various IoT techniques used in smart farming.

Chapter 2 aimed to survey different applications of wireless sensor networks (WSN) in healthcare. This chapter focused on the interoperability of sensor data to build promising and interoperable domain-specific or cross-domain IoT applications. Here, the authors described various challenges of healthcare applications, different platforms of the applications, and the security issues of the healthcare infrastructure.

Chapter 3 aimed to discuss the energy-efficient routing with mobile sink protocols that are more suitable to strengthen the agriculture. This chapter organized by classifying aforesaid protocols into three different categories (e.g., hierarchical-based, tree-based, and virtual-structure-based routing). The discussion concludes that packet congestion is minimized along with energy optimization among sensor nodes. So, the dynamic rearrangement of routing tree makes the system more efficient towards network delay and improving of network lifetime.

Chapter 4 explained the design of a capacitance sensor array for analyzing and imaging the non-homogeneity in biological materials. Further, tissue mimicking phantoms are developed using Agar-Agar and Polyacrylamide gels for testing the developed sensor. Also, the sensor employs an unsupervised learning algorithm for automated analysis of non-homogeneity. The reconstructed capacitance image can also be sensitive to topographical and morphological variations in the sample. The proposed method is further validated using a fiberoptic-based laser imaging system and the Jacquard index. In this chapter, the design of the sensor array for smart analysis of non-homogeneity along with significant results are presented in detail.

Chapter 5 deals with LEACH protocol, which is a cluster-based routing protocol. As multiple nodes are required for cooperative communication, the Low Energy Adaptive Clustering Hierarchy (LEACH) protocol is used for cluster formation. Further, vector quantization (VQ) is used for analysis of low energy path for the nodes and clusters respectively. In this chapter, the proposed technique illustrates the LEACH-Vector Quantization (LEACH-V) protocol, for intra-cluster communication in a cooperative communication network. LEACH provides the optimum cluster size and their cluster head and using VQ. The minimum distance is calculated using Euclidean distance between the multiple cluster heads which creates the shortest path results in energy efficient technique. The proposed work illustrates the LEACH-Vector Quantization Dijkstra (LEACH-VD) protocol, for shortest path active cluster head (CH) communication on a cooperative communication network. In the application point of view, LEACH-VD performs the lowest energy path. LEACH-V provides the intra-cluster communication between the cluster head and using Dijkstra Algorithm. The minimum distance is calculated connecting the active cluster heads, which creates the shortest path results using an energy-efficient technique.

Chapter 6 describes how agriculture is a sector that holds great promise to Indian economic growth. Production in rural Tamil Nadu is extremely low due to unscientific farming practices. The major challenges faced in Tamil Nadu agriculture are crop mapping, yield prediction, quality of food produced, irrigation management, variable rate fertilizer, and pesticide due to lack of technical knowledge. Precision agriculture (PA) rules out all drawbacks of traditional agriculture. The main objective of the chapter is to enhance the productivity of rural Tamil Nadu in order to meet the growing demands of our country's food supply chain.

Chapter 7 aimed to choose the best location for mustered oil mill location. Selecting the best location for setting up a mustard mill can be considered a multiple-criteria decision-making problem (MCDM), and ELECTRE III method is used and explained in detail to rank different location options in increasing order of suitability.

Finally, a conclusion of the book is given. The application of expert systems in various fields of agriculture technology has been discussed.

Proshikshya Mukherjee
KIIT University (Deemed), India

Prasant Kumar Pattnaik
KIIT University (Deemed), India

Surya Narayan Panda
Chitkara University, India

Acknowledgment

The editors would like to acknowledge the help of all the people involved in this project and, more specifically, to the authors and reviewers that took part in the review process. Without their support, this book would not have become a reality.

First, the editors would like to thank each one of the authors for their contributions. Our sincere gratitude goes to the chapter's authors who contributed their time and expertise to this book. They are all models of professionalism, responsiveness and patience with respect to my cheer leading and cajoling. The group efforts that created this book are much larger, deeper and of higher quality than any individual could have created. Each and every chapter in this book has been written by a carefully selected distinguished specialist, ensuring that the greatest depth of understanding be communicated to the readers. We have also taken time to read each and every word of every chapter and have provided extensive feedback to the chapter authors in seeking to make the book perfect. Owing primarily to their efforts I feel certain that this book will prove to be an essential and indispensable resource for years to come.

As editors would like to thank the Cloud Computing System and Agriculture technology community for recognizing the quality, effort and care that has been made (by many) in creating such a successful, wonderful and useful product.

Second, the editors wish to acknowledge the valuable contributions of the reviewers regarding the improvement of quality, coherence, and content presentation of chapters. Most of the authors also served as referees; we highly appreciate their double task.

Acknowledgment

There have been several influences from our family and friends who have sacrificed lot of their time and attention to ensure that we are kept motivated to complete this crucial project.

Last but not least the editors would like to thank all members of IGI Global publication for their timely help; constant inspiration and encouragement with friendly support to publish the book in time.

Proshikshya Mukherjee
KIIT University (Deemed), India

Prasant Kumar Pattnaik
KIIT University (Deemed), India

Surya Narayan Panda
Chitkara University, India

Introduction

In the 21st century, agriculture will go round into water-economy agriculture, intelligent agriculture, good-value high-acquiesce, and pollution-free agriculture. Agriculture information technology is a necessary to recognize all of purposes, and to utilize the digitization of each process in every aspect of agriculture through Information and Communication Technology. At some stage in this process of agriculture information, it is necessary to bring together all the components of the Agriculture digital system and handles different data types. At the present time, Agriculture Information Technology is progressively increasing in their level by incorporating new technologies with Wireless Sensor Network and Internet of Things. As a result of this, computers have been popularly adopted in farming, crop reproduction, management of crops and forest and insect identification, agricultural climate report, agricultural manufactured goods processing and so on. Again, Farmers are connected with the massive information flow, a farmhouse at home connected to informatics centers, varsities, R&D institutes in order to get updated information of cost, seed, new type of agricultural equipments and health care of plant so on. However, the environment of agriculture is a very dense natural, balance system and includes many factors, that is, from the environment to humanitarian, from ecology to trade and industry, from natural features to people, etc., that incurs costly technology.

In 1995, the concept about the connection of "thing to thing" was proposed by Bill Gates in the book *The Road Ahead*. After that in 1999, EPCglobal unified more than 100 enterprises and shaped ITU and formally internet of Things. The technology of IoT primarily focuses on the communication between thing to thing, human to thing, and human to human, ecology of human, plant to human so on with smart electronic devices via internet.

Wireless sensor network collects the data from different types of sensors and then sends it to the repository with information about diverse environmental factors that helps in monitoring as and when required. However in the context of agriculture, WSN may useful in the scenarios like attack of insects and vermin controlled by spraying with a suitable insecticide and pesticides.

There is a risk of theft when plants are at the phase of harvest. Smart GPS-based distant operated robots can weed, spray, sense wetness, scare birds and animals, etc. Secondly, it includes smart irrigation with smart control based on real-time field data. Thirdly, smart warehouse management can maintain temperature, maintain moisture, and identify theft. Managing all these operations will be through any digital device connected to the internet, and the operations will be controlled by interfacing sensors.

Proshikshya Mukherjee
KIIT University (Deemed), India

Prasant Kumar Pattnaik
KIIT University (Deemed), India

Surya Narayan Panda
Chitkara University, India

Chapter 1
IoT–Based Precision Agriculture System:
A Review

Sarita Tripathy
KIIT University (Deemed), India

Shaswati Patra
KIIT University (Deemed), India

ABSTRACT

The huge number of items associated with web is known as the internet of things. It is associated with worldwide data consisting of various components and different types of gadgets, sensors, and software, and a large variety of other instruments. A large number of applications that are required in the field of agriculture should implement methods that should be realistic and reliable. Precision agriculture practices in farming are more efficient than traditional farming techniques. Precision farming simultaneously analyzes data along with generating it by the use of sensors. The application areas include tracking of farm vehicles, monitoring of the livestock, observation of field, and monitoring of storage. This type of system is already being accepted and adopted in many countries. The modern method of smart farming has started utilizing the IoT for better and faster yield of crops. This chapter gives a review of the various IoT techniques used in smart farming.

DOI: 10.4018/978-1-5225-9004-0.ch001

INTRODUCTION

The basic source of livelihood in India is agriculture. Hence the survival of human species is based mainly on agriculture. The development of countries economy is dependent on it, as it is the most important source of food. Agriculture also provides ample opportunities of employment to the people. In present world farmers are still following the conventional methods of agriculture, the consequence is that there is low production of crops and fruits. This situation can be improved if we follow new techniques which utilize automatic machineries. As per the survey done so far there is no significant development in the agriculture sector. Due to decline in the crop rate food prices are continuously increasing. Due to this large number of people are pushed into poverty. Factors such as wastage of water, less fertile soil, inadequate supply of fertilizer, change in climatic condition, *etc*. At present it is essential for us to create solution for this problem by applying new techniques. The technologies such as IOT can make effective development in agriculture. What is the need of IoT in agriculture? (Rao and Sridhar, 2018). The survey of food and agriculture organizations-United Nations has given the data that around 70% growth in agricultural production is required till 2050 keeping in eye the evolving population. The increase in crop production can happen if the modern science and technology is applied. By the use of IoT, there can be increase in production along with efficient monitoring of soil at a minimum cost, monitoring of temperature and humidity, monitoring of rain fall, checking of efficiency of soil, water tank storage capacity monitoring and also detection of rate of theft. Modernization of agriculture can be done by combining the traditional methods with technologies such as IoT.

A three tier system is involved in internet of things (Patil and Kale, 2016). The three layers are perception, network and application. The main component present in the first layer are sensor motes which are the devices enabled with Information Communication Technology(ICT).The main building blocks of sensor technology are the sensor motes. The components included in it are RFID tags, sensor network recognizable objects and sensor objects which are able to collect real time information. The IoT infrastructure which realizes the universal space is the network layer. Then comes the two other layers which are the combination of application layer and perception layer. Any specific industry can be combined with IoT through the application layer. Different areas of industry which includes smart agriculture, smart parking, smart building environment monitoring, transportation and healthcare. Agriculture is certainly the most important area as millions of people are affected by this. Farmers in modern world are getting educated and having knowledge about new technologies. They are being educated about the internet and use of social websites through which they are getting regular updates, and they are able to collect huge amount of data through it.The connection of devices and data information collection are two important uses of internet of things. IoT infrastructure and frameworks are used to handle and also interpret the different types of data and information's in various fields. Specific sensors can be registered by the users and also they can process data and information and create streams. 'Internet of Things 'consists of devices which are capable of analysing and monitoring the information through sensors and also simultaneously transmit it to users, which are utilized in a variety of agricultural methodologies.

BACKGROUND

In Abhijith, Jain, and Rao, 2017, the authors have described how the process of agriculture can be advanced with the integration of internet of things with it. According to their work division of the entire farm is done into equal sized grids and the individual grids are equipped with sensors. The soil sensor sense the ph level and calcium content of the soil and environmental sensor sense the temperature, humidity and send it to the sink node. The collected data is send to the mobile application. The mobile apply some predictive algorithm and analyse whether the soil and environmental condition is suitable for the specified curb or not and send an alert message accordingly.Agriculture depends upon various underground things like soil moisture, salinity, and temperature of earth. In Vuran, Salam, Wong, Irmak, 2018, they have described a new internet of things i.e. internet of underground things. IOUT is an autonomous device that collects the data like soil moisture, salinity, and temperature of earth. These information are integrated with the field machinery like irrigation system, harvesters, and seeders to automate it.The farmer required different machinery at different phases of agriculture which may not be available on time. In Zhang, Hao, and Sun, 2017, the authors have proposed an agricultural machinery management system that takes the advantages of modern technologies like cloud computing and the internet of things. This system includes six layers the first layer is the sensor layer which gets all the information including the location, running status, working time of agricultural machinery. The network layer provides a channel to transport information between system and devices including GPRS, WIFI, Internet, and mobile communication system. The data service layer processes not only the data from the device but also the data from the company's ERP and OA system which is build on the cloud computing platform. The application layer completes the management of a lot of large scale agricultural machinery with the internet. The access layer includes PC, notebook, mobile phone and other smart devices which can run the application software. The user layer represents those users who can access the system. In Zhang, Hao, and Sun, 2017, the author have designed a remote monitoring system for agricultural field which provides alert massaging service, weather forecasting, crop conditions etc. Total three modules are present in the remote monitoring system they are the farm, service and the client sides. The farm side modules sense the agricultural parameters and send it to mobile application. The mobile application receives the data and compare with the data present in server and send the alert message to the farmer.

Climate condition is a major factor in agriculture, and the farmer performs different operation in the agricultural field accordingly. In Patil and Kale, 2016, the author has described how internet of things and artificial hydrocarbon network model can be used to predict the climate condition at a remote place. The IoT systems senses the temperature, humidity and pressure of a remote location and send the data to the artificial hydrocarbon network (AHN) model. The AHN is a supervised learning algorithm, which takes data for IoT system and test with the web server data and predicts the temperature accordingly. Starting from the cultivation to reach up-to the end users it under goes a series of process. The production environment, processes, testing etc. needs to be supervised and managed so that the quality of the product is maintained. In Martin-Garin, Millan-Garcia, Millan-Medel according to the authors, IoT can be used to maintain the quality, productivity and security of the product. To improve the agricultural product they have integrated the three networks given as the internet, the supply chain, IoT. The internet is required to connect the technical resources. The supply chain includes five links like supply of raw materials, production, consumption and processing circulation which are needed for maintenance of quality and safety of products that IoT uses. IoT can also be used to store the cereals safely in the godown. After production the cereals are stored in the godown from where it is being exported to different places. There is a chance of the stock to be stolen. In Vuran, Salam, Wong, and Irmak, 2018, the authors have introduced a smart tag tracking system to store the product safely. The smart tag tracking system consists of a GPS module which sends the position of the object i.e.

latitude, longitude and id of smart tag to the micro-controller. On receiving a new value the micro-controller decode it and send it to mobile application through 3G module. The mobile application compare the value received from 3G module with the value of database server and detect whether the object is at its correct location or not. IOT has tremendous application in the agricultural field. In Zhang, Hao, Sun, 2017, it is stated that the architecture of IoT consists of three numbers of layers they are the perception, network and application. The technologies used by the first layer are RFID, WSN, near field communication for receiving the data from the agricultural field. The network layer used to make the communication between the wireless sensor nodes and physical objects, and is also used to transfer the data to a remote infrastructure for storage. The application layer is used to identify the devices, collect data from all the devices, perform some computation and give some alert message to the user.

DISCUSSION

Integration of IoT in agricultural field has tremendous advantage. With the application of IoT the farmer can know the properties of soil, the climatic condition, probability of rain fall etc. The farmer can monitor his farming from a remote place with the use of smart devices. Farmer requires a lot of machinery for cultivation, which can also be managed with the help of IoT. For storing and supplying the crop smart tagging and supply chain management

Table 1.

Factors	Description
Cost	The IOT devices may not be affordable for small scale and marginal farmer. So the IOT device is cost effective.
User friendly	The farmer may be unaware of using smart phone. The mobile application used for IOT should be simple so that user can easily access the device.
Availability of resources	The resources like sensor, network facilities and smart phone used in IoT may not be available for all the farmer.
Accuracy	The data collected by the sensor and received by the application must be accurate so that the farmer can get the correct information and take the necessary action.
Security	The data communicated between the sensor and the mobile application must be secure and must not be corrupted during transmission.
Correct Prediction	The mobile applications are also used to predict the future climate condition by using machine learning algorithm. Proper algorithm must be used for prediction so that the farmer can get the correct information.

can also be designed with the help of IoT. The implementation of IoT in agricultural field requires sensors, internet facility and smart devices. There are certain factors that have to consider for implementing an IoT device.

CONCLUSION AND FUTURE SCOPE

IOT has the application in smart home, smart city, smart grid, agricultural system, health care, transportaïon etc. In agricultural field IoT is used to provide proper information about the crop, weather, properties of soil, availability of machinery, supply chain management, smart tagging system etc. According to the economic condition of the farmer all this facilities may not be affordable for small and marginal farmer. So the IoT devices should be cost effective, and easily available to the farmer. Future work will consist of smart and efficient system of security which would provide protection from outsiders into the field or any rodents or animals. The system would be able to recognize it and make the farmer aware of it so that appropriate steps can be taken immediately.

REFERENCES

Abhijith, H. V., Jain, D. A., & Athreya Rao, U. A. (2017). Intelligent agriculture mechanism using internet of things. *2017 International Conference on Advances in Computing, Communications and Informatics (ICACCI)*, 2185-2188. 10.1109/ICACCI.2017.8126169

Martín-Garín, Millán-García, Baïri, Millán-Medel, & Sala-Lizarraga. (2018). Environmental monitoring system based on an Open Source Platform and the Internet of Things for a building energy retrofit. *Automation in Construction*, *87*, 201–214.

Patil, K. A., & Kale, N. R. (2016). A model for smart agriculture using IoT. *2016 International Conference on Global Trends in Signal Processing, Information Computing and Communication (ICGTSPICC)*, 543-545. 10.1109/ICGTSPICC.2016.7955360

Rao, R. N., & Sridhar, B. (2018). IoT based smart crop-field monitoring and automation irrigation system. *2018 2nd International Conference on Inventive Systems and Control (ICISC)*, 478-483. 10.1109/ICISC.2018.8399118

Vuran, Salam, Wong, & Irmak. (2018). Internet of underground things in precision agriculture: Architecture and technology aspects. *Ad Hoc Networks, 81*, 160-173.

Zhang, R., Hao, F., & Sun, X. (2017). The Design of Agricultural Machinery Service Management System Based on Internet of Things. *Procedia Computer Science, 107*, 53–57.

Chapter 2
Applications of Wireless Sensor Networks in Healthcare

Ramgopal Kashyap
iD https://orcid.org/0000-0002-5352-1286
Amity University Chhattisgarh, India

ABSTRACT

Health is the key capability humans require to perceive, feel, and act effectively, and as such, it represents a primary element in the development of the individual and the environment humans belong to. It is necessary to provide adequate ways and means to ensure the appropriate healthcare delivery based on parameter monitoring and directly providing medical assistance. Wireless sensor networks (WSNs), commonly known as the internet of things (IoT), enable a global approach to the healthcare system infrastructure development. This leads to an e-health system that, in real time, supplies a valuable set of information relevant to all of the stakeholders regardless of their current location. Economic systems in this area usually do not meet the general patient needs, and those that do are usually economically unacceptable due to the high operational and development costs. This chapter shows how recent advances in wireless networks and electronics have led to the emergence of WSNs in healthcare.

DOI: 10.4018/978-1-5225-9004-0.ch002

INTRODUCTION

WSNs consist of spatially distributed autonomous devices to cooperatively monitor real-world physical or environmental conditions, such as temperature, sound, vibration, pressure, motion, pollution and location. This technology is also widely used by military applications, such as battlefield surveillance, transportation monitoring, and sensing of nuclear, biological and chemical agents. Recently, this technology has developed and been widely used in daily life as WSNs are low cost, low power, rapid deployment, have self-organizing capability and cooperative data processing, including applications for habitat monitoring, intelligent agriculture and home automation. The major components of a normal WSN sensor node are a microcontroller, memory, transceiver, power source and one or more sensors to detect the physical phenomena. The structure of the sensor node is generally divided into four major parts: sensing unit, processing unit, communication unit and power unit (Hejlová & Voženílek, 2013). A sensor node sends the measurement of the physical phenomenon to the sink which has bigger memory and processing power. Depending on the application scenario, sometimes extra hardware is added in the sensor nodes and a deployment strategy is devised. Normally, in applications for WSNs the environment is unpredictable such as hostile, with remote harsh fields or disaster areas, sometimes called toxic environments. Hence, no standard deployment strategy existed the deployment usually involves scattering or by possibly carrying out the application scenario (Brinis & Saidane, 2016). Despite their quick deployment and significant advantages over traditional methods, WSNs have to face various security problems because of their nature and the possibility of the presence of one or more faulty or malicious nodes in the existing network. There are many technically interesting research discussions involving WSNs, such as development of models and tools for the design of better WSNs architecture and elaboration of standard protocols in WSN adapted to work robustly on certain scenarios. However, one of the most important issues that remain subject to debate is security. The emphasis in this chapter focuses on security in WSNs. More precisely, the work focuses on investigating models preventing internal attacks on WSNs.

Internet of Things

The ability of everyday devices to communicate with each other and/or with humans is becoming more prevalent and often is referred to as the Internet of Things (IoT). It is a highly dynamic and radically distributed networked system, composed of a very large number of smart objects. Three main system-level characteristics of IoT are: Anything communicates; anything is identified; and anything interacts. The basic and most important smart object of IoT is a sensor node, or more precisely, Sensor Web (SW), which can be defined as a web of interconnected heterogeneous sensors that are interoperable, intelligent, dynamic, scalable and flexible (Zhang, Nagarajan & Nevat, 2017). These smart objects are envisaged to provide smart metering, e-health logistics, building and home automation and many new uses not yet defined. The human health today presents the primary focus of an increasing number of case studies and projects whose goal is healthcare improvements and achieving the foundations for a global health system. Such systems should provide information to the patients and their doctors, regardless of the location where they are located. This is known as e-health, and today is closely related to the Internet including and applying IoT concepts to such defined global system, the possibility of its application for saving lives becomes unlimited.

Various e-health scenarios are enabled by rapid advancements in information and communications (IC) technologies and with the increasing number of smart things (portable devices and sensors). IoT powered e-health solutions provide a great wealth of information that can be used to make actionable decisions. By connecting information, people, devices, processes and context, IoT powered e-health creates a lot of opportunities to improve outcomes (Li, Jin, Yuan & Zhang, 2018). Some of the most promising use cases of connected e-health include preventive health, proactive monitoring, follow-up care and chronic care disease management. It can be stated that the IoT is a disruptive innovation, which bridges interoperability challenges to radically change the way in which healthcare will be delivered, driving better outcomes, increasing efficiency, and making healthcare affordable. In order to track and record personal data it is necessary to use sensors or tools which are readily available to the general public. Such sensors are usually wearable devices and the tools are digitally available through mobile device applications. These self-monitoring devices are created for the purpose of allowing personal data to be instantly available to the individual to be analyzed. The biggest benefit of self-monitoring devices is the elimination of the necessity for third-party hospitals to run tests, which are both expensive and lengthy. These devices are an important advancement in the field of personal health management. Currently, self-monitoring healthcare devices exist in many forms this chapter presents the creation of such system which will include everything user would expect from a commercial system. In other words, by using inexpensive hardware and open source software, it is possible to create a DIY (Do-It-Yourself) system in such a way that own solution providing techniques to end-users and the possibility to shape products according to their needs is beneficial for both users and product developers. WSN for health care systems transmit a large volume of data collected from several bodily sensors. Given that the information regarding the health of an individual is highly sensitive and critical, it must be kept private and secure. Consequently, it is important to equip the system with mechanisms that would prevent unauthorized agents from acquiring or tampering sensitive data that is transmitted over the network and stored in dedicated repositories.

The aim of this chapter is to provide an Internet of Things based healthcare information system intended for indoor and outdoor use where a methodological approach to the design process is in focus. A distinguishing feature of this approach is that the contracting authority's and future users' perspectives and needs are included in most stages of the design process. Moreover, in the approach, the designer from the beginning has to think comprehensively to merge human and technical constraints and requirements. The user-driven Design Methodology (DM) is used to solve the problems of the real-life scenario of supporting seniors living alone, especially those with limited abilities to manage their daily lives (Watanabe, Fukuda & Nishimura, 2015). The conducted design process results in a system proposal that meets the required assumptions. The conducted case studies verified that the designed system, consisting of the Inertial Measurement Unit (IMU) with a built-in three-axis accelerometer, gyroscope, magnetometer and altimeter, together with Wi-Fi and heart rate modules and applying thresholding, Pedestrian Dead Reckoning (PDR) and decision trees algorithms, works properly in the tested real environment (Kuang, Niu & Chen, 2018). The achieved person's localization accuracy within one meter fits the required four room-zone level localization accuracy in an apartment environment. The developed fall detection algorithm proved effectiveness of 98%, and other required activities were recognized with 95% compliance. Moreover, the behavior classification algorithm is able to distinguish normal behaviors from suspicious and dangerous ones, working properly in almost 100% of cases.

Motivation

Wireless communication is the transfer of information between two or more points that are not connected by electrical conductors. Most of the wireless communication technology uses radio waves in order to transfer information between the points which are known as nodes. One application domain of wireless communication is wireless sensor networks. WSN is a distributed system, containing resource or constrained nodes that work in an ad hoc manner using multi-hop communication. WSNs and Internet are integrated as a new application area called IoT, covering almost every area in current daily life (Xu, Collier & O'Hare, 2017). IoT encourages several novel and existing applications such as environment monitoring, infrastructure management, public safety, medical and health care, home and office security, transportation, and military applications. Figure 1 shows the complexity of wireless sensor networks, which translate sensing and identification activities into services using WSNs with WSN middleware and access networking. It can use: (i) different communication platforms such as WiFi, wireless LAN, 3G and 4G; (ii) different devices which are based on different processors such as various types of PDA, smartphones and laptops and (iii) all these platforms and devices being built on different architectures such as centralized, distributed or peer-to-peer. WSNs provide unprecedented ability to identify, observe and understand large-scale, real-world phenomena at a fine spatial-temporal resolution. The applications range from military to daily life. For example, in community services WSNs can (1) provide early warnings for natural disasters such as floods, hurricanes, droughts, earthquakes, epidemics; (2) disseminate surveillance information for cities in parks, hotels, forests, to support municipality service delivery; and (3) provide enjoyment of the city by citizens and tourists through public services support such as monitoring of water quality to ensure that citizens always have clean water or providing free environmental information on the main tourist destinations. In general, the network consists of a data acquisition network and a data distribution network, monitored and controlled by a management centre. Security is an inevitable need both in wired and wireless communication networks. The ultimate security aim in both networks is to provide confidentiality, integrity, authenticity, and availability of all messages in the presence of resourceful adversaries (Aminzade, 2018). Every eligible receiver should receive all messages intended for the message recipient and be able to verify the integrity

Figure 1. The complexity of wireless sensor network

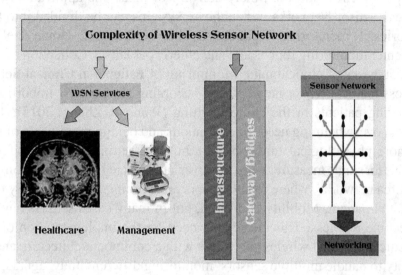

of every message as well as the identity of the sender. Adversaries should not be able to infer the contents of any message.

In the wireless paradigm wireless sensor networks continue to grow because of their application scenarios and cost effectiveness. A major benefit of these systems is that they perform in-network processing to reduce large streams of raw data into useful aggregated information. Protecting information is critical. The traditional computer network security goal is to deliver the message to the end user in a reliable way. The leading traffic pattern in the conventional computer network is end to end communication. The message content is not important beyond the necessary header (Monika & Upadhyaya, 2015). In this process the message authenticity, integrity and confidentiality are usually achieved by an end to end security mechanism such as Secure Socket Layer (SSL).

WIRELESS SENSOR NETWORKS

Wireless sensor networks have emerged as a feasible technology for a myriad of applications, including many different health care applications. WSN technology can be adapted for the design of practical Health Care WSNs (HCWSNs) that support the key system architecture requirements of reliable communication, node mobility support, multicast technology, energy efficiency, and the timely delivery of data. The application of the Wireless Sensor Networks in healthcare systems can be divided into three categories: 1. monitoring of patients in clinical settings 2. Home & elderly care center monitoring for chronic and elderly patients 3. Collection of long-term databases of clinical data. Monitoring of Patients in Clinical Settings Wireless medical sensor networks are becoming increasingly important for monitoring patients in the clinical setting (Wang & Zhang, 2015). There exists an overwhelming need for continuous and benign monitoring of more and more physiological functions in a hospital setting. Sensors today are effective for single measurements, however, are not integrated into a "complete body area network", where many sensors are working simultaneously on an individual patient. Mobility is desired, but in many cases sensors have not yet become wireless. This creates the need for the implementation of new biomedical personal wireless networks with a common architecture and the capacity to handle multiple sensors, monitoring different body signals, with different requirements. The type and number of sensors must be configured according to monitoring needs related to different diseases, treatment, and the patient treatment life cycle. WMSNs systems have several advantages over traditional wired systems such as ease of use, reduced risk of infections, reduced risk of failures, reduced user discomfort, enhanced mobility, increasing the efficiency of treatment at hospital, and lower cost of delivery.

Internet of Everything

Internet of Everything (IoE) contacts various parts of life, just a few cases fuse related homes and urban zones, related automobiles and roads, or devices that track a man's lead and use the data accumulated for "push" organizations. Different cases consolidate therapeutic sensors that enable tweaked e-Health, and sensors that screen current equipment to increase use in collecting. Uses of IOE run from the gimmicky to the beguilement changing, from advanced cell empowered ease back cookers to trackers which prepared authorities when anchored forests tumble to unlawful logging. Regardless, the significance of IOE, generally called the web of things or the advanced web, is more than singular things. It is the canny relationship of people, methodology, data, and things. IOE incorporates three parts: the physical thing itself, sharp fragments, for instance, sensors and programming, and accessibility with a more broad structure, either between singular things, among things and a central framework between various things and outside data sources. The blend is more than the entire of its parts; structures, for example, combine commitments from control customers in homes and work environments with commitments from control providers in plants and twist ranches to choose when and where imperativeness is used, changing cost to reflect ask for and finally decreasing general essentialness use. The fast improvement in data creation and transmission, in any case, only a little bit of every single physical inquiry on the planet are starting at now connected with IP frameworks (Snyder & Byrd, 2017). Cisco assesses that under 1 percent of physical things are related with IP arranges however the IoE is reaching out as more contraptions and customers are interfacing with IP sorts out every day, driving more trades and techniques on the web. For individuals, the impacts of the IoE are felt step by step. Sensors embedded in shoes, for example, track the divisions that wellbeing aficionados run and subsequently exchange information to web-based systems administration profiles to rapidly differentiate athletic achievements and those of sidekicks. Web engaged wake-up clocks amass data on atmosphere and development, uniting that information with a customer's logbook, choosing the perfect time to wake neighborhood occupants. In addition, applications on cutting edge cell phones can control home electronic devices, changing warming and cooling levels and furthermore preparing security settings remotely (Papers, 2018). At a cutting edge level, applications using sensor headways are getting gigantic measures of data to improve fundamental initiative. Sensors introduced in plant fields screen temperature and soddenness levels, controlling water framework systems. Contraptions in oil fields moreover, significant well mechanical assemblies track all parts of infiltrating and fuel movement, growing age efficiency. Moreover, sensors in vehicles can screen usage, lighting up decisions around refueling and repair and furthermore vehicle plot. For governments, IoE and gigantic data applications are checking pandemics and common conditions improve open prosperity and security, and augmentation viability in the transport of open organizations, for example, common development structures that wire steady remote checking to streamline movement streams.

Solid Pillars of IoE

There are four mainstays of IoE: People, Process, Data and Things extended accessibility can enhance each bit of the IoE star gathering, from work process efficiencies to quiet satisfaction and results.

Individuals: The accessibility given by IoE improves the idea of solution for everyone included. Patients are spared from giving monotonous data at different reasons for area. Their recuperating office encounter is enhanced through chronicles that illuminate more about their affliction or condition, expanding their base of data. Test outcomes are more provoke. Their prosperity history is accessible to each boss in their line of care, empowering correspondence and composed exertion. Basic information is passed on to the patient through various reasons for relationship from wearables, tablets or propelled cells. Undoubtedly, even seemingly insignificant details, like the ability to ask for sustenance direct from the dietician, improve the patient's contribution and satisfaction scores as shown in figure 2.

Process: New development is basically especially to address the issue of waste in therapeutic work process. Testing gives an average representation. Consistently test occurs remain new until the point that the therapeutic guardian or pro has space plan astute to check their fax or inbox. Information can be examined at the Internet edge, sorted out by alert level and sent clearly to a parental figure to make a move as required (Papers, 2018). A broad number of assignments are holding up to be automated endeavors that will save time, diminish expenses and improve calm outcomes. As patients are diverted to more appropriate personality settings, weight on the triage staff is remarkably

Figure 2. Strong pillars of internet of everything

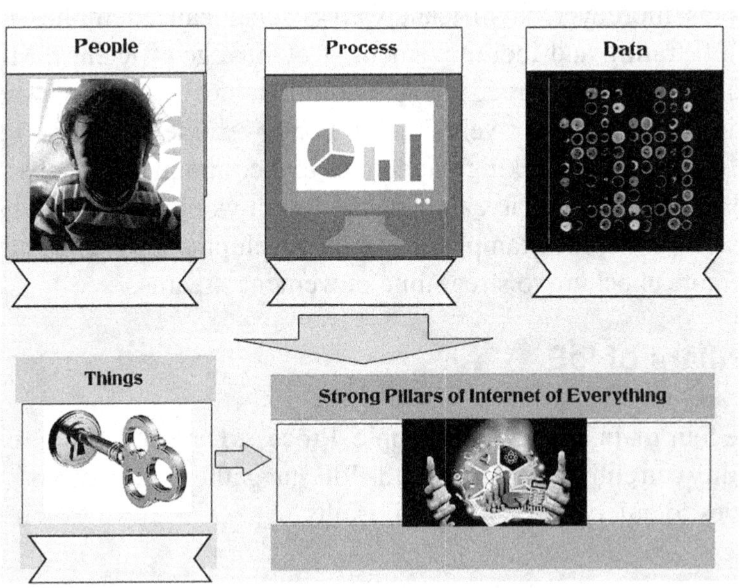

diminished. Turning away silly certifications maintains a strategic distance from misuse, while it redesigns staff time for those most in require.

Data: This area of IoE holds the best assurance for setting up upgraded methodologies for social protection movement. Over 78% of prosperity providers nitty gritty they were gathering electronic data, despite the way that the information has a tendency to sit away unexploited. A conventional social protection head has distinctive applications that assemble and extra information anyway little is done to change over that data into critical encounters for a pro, sustain or other provider (Kashyap and Gautam, 2017). Endeavoring to vanquish these obstructions, Cisco is making it possible to securely facilitate and sever down data without exchanging tolerant protection. Information imperative to providers as appeared in figure 3, it can be a tick a long way from aiding in a care decision. Think about the vitality of promptly differentiating a patient's history and those in equivalent conditions. In a moment acknowledge what worked, what didn't.

Think about it from a patient's perspective, especially in the occasion that they've changed starting with one specialist then onto the next rather than sit tight for a patient to recuperate printed duplicates of X-beams or diverse films, imagine that they are open instantly through the Internet. Cisco's ensured affiliations are lessening downtime as they strengthen remedial fundamental administration limits.

Figure 3. Challenges in the future

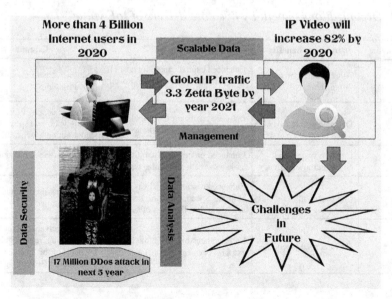

Things: In the Internet of Everything (IoE) things, are motivations behind affiliation. In the space of medicine these range from cutting edge cell phones and PCs to wearables and other complex helpful devices. The building that offers learning to these things is the Cisco compose information is made by requesting calm information, recognizing issues, and advised parental assumes that healing movement is required.

Eventual Fate of IoE for Society

As billions and even trillions of sensors are set the world over and in our atmosphere, we will get the ability to really hear our existence's "heartbeat." Indeed, we will know when our planet is strong or wiped out. With this suggests understanding, we can begin to execute a portion of our for the most part pressing difficulties, including hunger and ensuring the openness of drinkable water. Craving: By appreciation and envisioning whole deal atmosphere outlines, agriculturists will have the ability to plant trims that have the absolute best for accomplishment (Condon, 2018).

Likewise, once the fields are harvested, more powerful (and, as needs be, more reasonable) transportation structures will think about the allocation and movement of sustenance from places where there is riches to spots where there is lack. Drinkable Water: While IoE will be notable make water where it is required most, it will have the capacity to settle a robust part of

Table 1. Organizations benefit by five core drivers of IoE value

S.N.	Industry Benefits	Role of IoE	Example
1	Asset benefit and cost diminishments	Enhanced capital capability cost of items diminishments and improved business plan execution	Shrewd structures, splendid creation lines
2.	Delegate productivity	Enhanced work profitability less or more gainful individual hours	Possible destiny of work, bring your own specific contraption", Versatility
3.	Creation system and coordination's adequacy	Enhanced process profitability lessened creation organize waste	Savvy arrange essentialness capability
4.	Improvement	Enhanced research, headway, and building, speed, New designs of activity and wellsprings of salary	Higher nature of things and organizations
5.	Customer Satisfaction	Enhanced customer lifetime regard extra bit of the pie (more customers)	Related advancing and publicizing, related guideline Effects of the change of the Internet of Everything

the issues that reduce our perfect water supply, for instance, mechanical waste, unsustainable agribusiness, and poor urban organizing. For example, splendid sensors arranged all through a city's water system will recognize when there is an opening and normally divert water to avoid pointless waste. A comparable sensor will prepared utility personnel with the goal that the issue can be settled when resources are open. Associations benefit by five fundamental drivers of IoE Value is appeared in table 1.

Business Strategies

The move from thing provoked organizations drove affiliations is one of the movements this move in business method has group recommendations for an open or private-division affiliation. Open and private fragment affiliations have started getting device driven game plans or Internet of everything answers for drive business handle efficiencies. Affiliations get these IoE courses of action whether clear sensors to screen fabricating plant rigging or applications to screen task force vehicles to help control working or capital costs (Papers, 2018). We trust affiliations that start using IoE answers for cost control will expand and overhaul their responses for support progression. This advancement will drive thing driven relationship to expand their portfolios to join more organizations. The IoE is one of the enabling advances in the improvement of the IoE world, and by abusing; it will help relationship of grouped sorts to see the broader plan of chances IoE affiliations make possible that is appeared in figure 4, how internet will change the lives.

Human Interaction With Technology

IoE courses of action increase levels of correspondence amongst individuals and advancement this extension in participation gets from the way IoE development is used. There is an exponentially extending estimation of IoE game plans as the human-to-advancement commitment exhibits increases and the explanation behind the IoE course of action gives more opportunities to one-way and two-course collaboration amongst things and people. See Figure 4. Underneath we depict four sorts of advanced courses of action and demonstrate the impact on human-to-development association.

Figure 4. Internet of Everything's
(Kashyap R, and Pearson A., 2018a)

Development Adoption

The change of the IoE affect influences advancement, Systems/Connectivity IoE courses of action will rely upon hybrid frameworks and system. Various open and private-fragment affiliations will pass on game plans that adventure of frameworks with expanded SLAs, security and extension characteristics that fit the necessities of the IoE game plan. Most related device courses of action today rely upon either settled or flexible frameworks for accessibility with little ability to pick the best framework over a given working period. This nonattendance of versatility brings about an answer with bring down general quality. IoE courses of action will misuse cream frameworks organization approaches that fuse 3G/4G/5G remote, Ethernet, diverse short-go arrange contributions, satellite correspondences and others.

Why IoE Is Important to You

The IoE raises your remedial I.Q. by giving you the right information, at the perfect time and in the advantageous place. In any case, it engages you to streamline your work procedure. Clear things, for instance, wiping out time spent get-together monotonous calm information can in a general sense cut down your cost of organization, free staff to perform distinctive commitments and give greater quality patient/master time. Redesigned calm outcomes and high patient satisfaction scores would increment be able to wage by raising your level of outcast portions course of action advancement is changing things. Once made only out of mechanical and electrical parts, things have ended up being capricious systems that unite gear, sensors, data storing, microchips, programming, and accessibility in load ways. These "keen, related things" rolled out possible by tremendous improvements in getting ready power and device downsizing and by the framework points of interest of inescapable remote accessibility have released some other time of competition (Condon, 2018). The changing method for things is furthermore disquieting quality chains, convincing associations to reexamine and retool all that they do inside. These new sorts of things change industry structure and the method for competition, introducing associations to new forceful open entryways and perils. They are reshaping industry breaking points and making absolutely new endeavors. In numerous associations, astute, related things will oblige the fundamental inquiry, "What business am I in?" related things raise another course of action of indispensable choices related to how regard is made and gotten, how the gigantic measure of new data they make is utilized and regulated, how relationship with ordinary business assistants, for instance, channels are rethought, and what part associations should play as industry limits are broadened. The articulation "Web of Everything's" has risen to reflect the creating number of splendid, related things and feature the new open entryways they can address (Ghoneim & Hussain, 2015). Anyway this articulation isn't astoundingly strong in understanding the wonder or its recommendations. The web, paying little mind to in the case of including people or things, is essentially a framework for transmitting information. What makes sagacious, related things basically unprecedented isn't the web, but instead the changing method for the "things." It is the expanded capacities of splendid, related things and the data they deliver that are presenting another time of competition. Associations must look past the progressions themselves to the engaged change happening. This structure engages remarkable new thing capacities. In the first place, things can screen and give their own one of a kind record condition and condition, making in advance difficult to reach bits of information into their execution and use. Second, complex thing activities can be controlled by the customers, through different remote get to decisions that gives customers the exceptional ability to alter the limit, execution, and interface of things and to work them in hazardous or hard to accomplish conditions. Third, the blend of checking data and remote-control limit makes new open entryways for upgrade. Figurings can liberally improve thing execution, utilize, and uptime, and how things work with related things in more broad systems, for instance, splendid structures and canny farms (Kashyap, Gautam & Tiwari, 2018). Fourth, the blend of checking data, remote control, and streamlining computations licenses independence. Things can learn, change in accordance with nature and to customer tendencies, advantage themselves, and work isolated.

THE IOT AND E-HEALTH

To improve human health and well-being is the ultimate goal of any economic, technological and social development. The concept of the IoT entails the

Figure 5. Confluence brought by the IoT

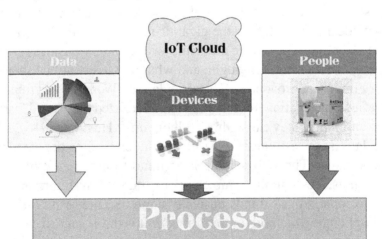

use of electronic devices that capture or monitor data and are connected to a private or public cloud, enabling them to automatically trigger certain field of healthcare. McKinsey Global Institute in its report presents predictions and economic feasibility of IoT powered healthcare, which states that by 2025 the largest percentage of the IoT incomes will go to healthcare as shown in Figure 5. Internet-connected devices, introduced to patients in various forms, enable tracking health information what is vital for some patients (Kashyap & Tiwari, 2018). This creates an opening for smarter devices to deliver more valuable data, lessening the need for direct patient-healthcare professional interaction (Ghoneim & Hussain, 2015). With faster, better insights, providers can improve patient care, chronic disease management, hospital administration and supply chain efficiencies, and provide medical services to more people at reduced costs. The IoT has already brought in significant changes in many areas of healthcare. It is rapidly changing the healthcare scenario by focusing on the way people; devices and applications are connected and interact with each other (Figure 5). Hence, it can be concluded that the emerging technology breakthrough of the IoT will offer promising solutions for healthcare, creating a more revolutionary archetype for healthcare industry developed on a privacy/security model.

Typically, IoT powered e-health solution includes the following functions: Tracking and monitoring (e.g. patient monitoring, chronic disease self-care, elderly persons monitoring or wellness and preventive care); Remote service; Information management; Cross-organization integration (Abdelwahab, Hamdaoui, Guizani & Rayes, 2014). The requirements of IoT communication framework in e-health applications are: Interoperability is needed to enable different things to cooperate in order to provide the desired service. Bounded latency and reliability are needed to be granted when dealing with emergency situations in order for the intervention to be effective. Authentication, privacy, and integrity are mandatory when sensitive data are exchanged across the network. A different technologies and architectures of IoT for healthcare can be found but next building elements are common for all of them: Sensors that collect data (medical sensors attached with the patient to measure vital parameters, and the environmental sensors which monitor the surroundings of the patient); Microcontrollers that process, analyze and wirelessly communicate the data; Microprocessors that enable rich graphical user interfaces; and Healthcare-specific gateways through which sensor data is further analyzed and sent to the cloud. In the next chapter a cheap solution for building IoT healthcare sensing devices will be presented.

Self-Monitoring Device

The main application domain of body sensor network (BSN) a network of sensors attached to the human bodies is continuous health monitoring and logging vital parameters of patients. In other words, these networked systems continuously monitor patients' physiological and physical conditions, and transmit sensed data in real time via either wired or wireless technology to a centralized location where the data can be monitored and processed by trained medical personnel. Medical devices must use a variety of design techniques to protect the underlying design as well as protect the sensitive data stored within or transmitted to/from the device. Many personal health monitoring devices must also be portable, so they need to be small, lightweight, and low-power (Peng, 2015). In order to build own health monitoring device with aforementioned characteristics, this chapter proposes a usage of ultra-cheap-yet-serviceable, small and powerful computer board - Raspberry Pi (RPi). A RPi has built in support for a large number of input and output peripherals and network communication, and it is the perfect platform for interfacing with many different devices and using in a wide range of applications. An intelligent home monitoring system based on ZigBee wireless sensors to assist and monitor the elderly people. The performance of their developed system was evaluated by running the system at four different elderly houses and recording the data and simultaneously performing the activity recognition in real time. The houses were equipped with the wireless sensor network with the fabricated sensor units attached to various house-hold appliances (Chen, 2016). Six electrical sensors are connected to appliances Microwave, Toaster, Water Kettle, Room Heater, TV and Audio. Four force sensors are connected to Bed, Couch, Dining chair and Toilet. One contact sensor is connected to grooming table and one temperature and humidity sensor to monitor the ambient environment readings. A laptop was installed with the developed intelligent software connected with ZigBee module acted as the coordinator associated with WSN to collect and monitor the elder's behavior. Growth of the elder population has given rise to the increasing demand for long-term elder-care facilities. The majority of elders living in such facilities are suffering from multiple chronic illnesses. Due to these illnesses, the elderly physical and mental health is slowly declining. In order to detect these declines early, continuous tracking of the elders' daily activity is required and integrated a 45-node WSN-based location system in NTUH-BH to automatically track the elder's daily mobility (Chao & Hsiao, 2014). They collected location traces and investigated the daily and long-term mobility of four volunteering elders for eight months. Following are the findings of their study: "Each elder's daily mobility shows a reoccurring pattern. The pattern, however, differs from individual to individual. The mobility level, the total distance the elder moves in the facility per day, shows a stronger variability. Not all elders show reoccurring patterns in mobility levels. Some of them are spontaneous moving from event to event." From these observations, it was concluded that long term location tracking, not just the mere quantity of mobility, allows discovery of the moving patterns and in turn making early detection of the elders' physical or mental problems possible. The web-based system was responsible for inferring high-order information about the activities of the condo's occupant and supporting the visualization in 2D Geographic Information System (GIS) and a 3D virtual world. They believe that one of the most innovative aspects of their project is the use of virtual world, a virtual world platform, for visualizing the activities of the condo tenant through an avatar. Only 9% of physicians work in rural areas, however, almost 20% of the population of the US lives in rural areas. There is a big shortage of physicians and specialist in the rural areas. Wireless Medical Sensor Networks technology has the potential to alleviate problematic patient access issues. At-home care can provide many advantages in terms of financial benefits, improved quality of life for patients, and more effective fall detection prevention or monitoring of many long-term chronic diseases (Domingo, 2014). Collection of Long-Term Databases of Clinical Data Sensors links the physical with digital world by capturing and revealing real-world phenomena and converting these into a form that can be processed, stored, and acted upon. The data that is gathered by the sensors in a WMSN can be used in two ways: 1. Healthcare applications that leverage wireless sensor networks analyze the data gathered by sensors to infer and make decisions about the state of a patient's health and wellbeing by improvement in monitoring consistency, continuous monitoring enhances data quality and precision for decision support leading to better titration of therapeutic interventions. 2. The continuous gathered data can be analyzed utilizing computational intelligence techniques to find solutions to the unsolved problems in the healthcare system.

Security and Privacy Concerns of IoT Powered E-Health

The realization of the IoT generally requires dramatic changes in systems, architectures and communications which should be flexible, adaptive,

Figure 6. Security challenges in IoT

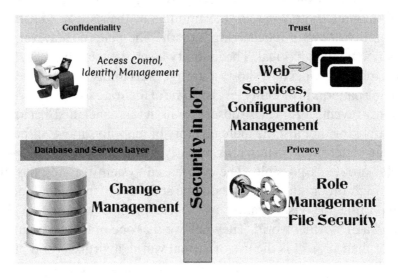

secure, and pervasive without being intrusive. The experts anticipate that the technologies of the IoT will bring forward critical developments in the areas of ethics, data protection, technical architecture, standards, identification of networked objects and governance. In other words, as more intelligent devices are connected to the Internet, the potential privacy implications and general false sense of security associated with weak key management and data compromise becomes critical (Scott & Mars, 2013). Thus, security (protection of data and privacy) represents a critical component for enabling the widespread adoption of IoT technologies and applications (Figure 6).

WSN Health Care System

Research group in Jönköping University has been research on the WSN health care system field for several years and has made remarkable achievement and great contribution on it. This group has established a relatively comprehensive WSN Health Care System theory and has developed a prototype to do experiment on it. Our gateway designs as part of Health care system base on the existing WSN Health Care System prototype in Jönköping University. See figure 7. This prototype consists of three parts.

The first part is monitoring sensor networks. It consisted of body sensor networks and home sensor network (HSN). BSN are sensor nodes attached on old people's body and provide the on body sensing information of him or her.

Figure 7. Component setup of the wireless health care system

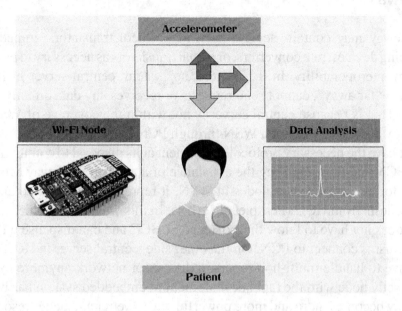

HSN are group of location-fixed sensor nodes with multiple sensors providing temperature, relative humidity, light sensors, microphone, and so on. HSN is distributed in living room, bedroom, kitchen, bathroom and corridor hiding in the sofa, bed or chairs. The second part is the access devices, which mean gateway. This gateway is usually a base station of the home sensor networks, which provide an interconnection between the monitoring sensor networks and central server system via Public Communication Network (PCN). It has to handle the connection problem between multi network communications. The third part is central server. It can be divided into four sub-parts: a conceptual database, a decision mechanism, a smart application gateway and a service management platform. The conceptual database is used to store the profiles information of the elders, the normal data collected by the sensor systems, the detection result, and report or alarm logs. Decision mechanism use Hidden Markov Model to detect the elder's coming action and health state. Once the dangerous situation is detected, the decision mechanism will drive the smart gateway to issue an alarm to the relatives to report the emergency situation (Morain, Kass & Grossmann, 2016). The service management platform is an interface of the whole system. The smart application gateway can establish communication with the caregivers to report the situation of the elders.

Gateway

A gateway may contain devices such as protocol translators, impedance matching devices, rate converters, or signal translators as necessary to provide system interoperability. In the health care system, central server in HCC, which is "far away", cannot send requests or receives any data directly from WSN. That is because central server is not within the coverage of WSN, It can only access services of a WSN through PCN but sensors nodes in WSN do not have the necessary protocol component at its disposal to communicate with PCN and thus requiring the assistance of a gateway. Gateway acts as a proxy for the set of sensor nodes in WSN. It represents all WSN to answer the requests from HCC (Pesko et al., 2014). In this way, on one side, sensor node does not have to know the existence of HCC and can dispense with all mechanisms connect to PCN. On the other side, central server in HCC does not have to handle multi-hop routes in the sensor network anymore to find the exactly node with the rapid development of embedded system hardware, gateway becomes more and more powerful due to better and better resources e.g. faster control unit and bigger memory. These make it possible that some tasks which originally belong to the central server can be transferred to the gateway. In this way, we can move the processing intelligence closer to the sensing device.

System Overview

Base on the system development research method, it is important to investigate the functionality and requirements of the smart gateway at the beginning. On one hand, the requirements analysis takes the system specification and project planning as a basic starting point of activities analysis, and performs inspection and adjustment from the software point of view; on the other hand, the requirements specification is the main basis of software design, implementation, testing and maintenance. The WSN health care system architecture is illustrated in figure 8. The whole system consists of four parts. The first part is the monitoring object; means elder's home where sensor nodes are probed to get multiple data of old people likes behave, activity, health state, and living environment information. Health care system has many old people, so this monitoring object is a plural. The second part is the monitor healthcare center, main job of healthcare center is to monitor health state of all old people in this system and make sure the normal operation of the

Figure 8. WSN health care system architecture

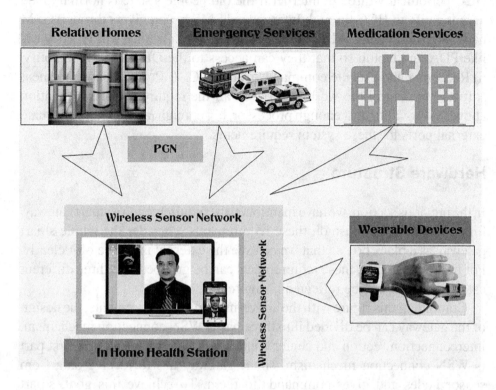

entire system (Dogan, 2016). Center server is located there to save necessary information of the elderly and provides variety of monitoring methods to indicate the current situation of the elder. Third part is care-givers including doctors or nurses in the hospital and the old people's relatives. They are responsible for dealing with the report message (normal or alarm message, from internet or SMS) that sent to them. They can also check the elder's current state through webpage which provide by Healthcare care. The fourth part is PCN including Internet, GSM/GPRS, Ethernet, and WI-FI. PCN connects all the other three parts together.

From this figure, we can see that gateway belong to part one (monitoring object), and it acts as bridge between WSN and PCN, offering multi communication ways to make sure required message can be transmit to desire destination. Detail requirements of smart gateway design are as follow: 1) Providing a mechanism to connect the WSN base station (sink) and received sensor collection data from it or send command to WSN. 2) Providing DB to save sensor data. 3) According to the current input sensor data and data stored in DB, determine the current health status of the elderly. 4) Sending notify to

HCC periodically through internet if the old people's state is normal. Send urgent notify to HCC through Internet and SMS to caregivers through GSM network. 5) Providing WLAN connection for short distance wireless devices like PDA and laptop so that they can access to the DB and receive notify. 6) Receiving and execute commands from HCC. 7) Providing management software to control the working flow of all the requirements that mention above. 8) Providing fast enough processor, big enough memory and adequate external ports for these system requirements.

Hardware Structure

In the previous section, we have mentioned the requirements of smart gateway. First thing to do is, base on these requirements, consider the whole smart gateway as a black box so that we can see the external interface of it clearly. For hardware design, these requirements can be grouped into three different categories. This can be reflected in figure 9.

Combining this figure with the above-mentioned requirements, the design of the gateway can be divided into three parts. WSN connection mechanism, interconnection section and center control unit. See figure 3-5 the first part is WSN connection mechanism, smart gateway has to receive data from sensor nodes and gives commands to them. To achieve this goal, smart

Figure 9. Smart gateway architecture

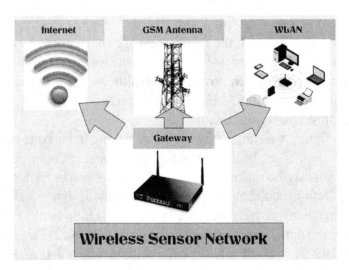

Figure 10. Intelligent model architecture

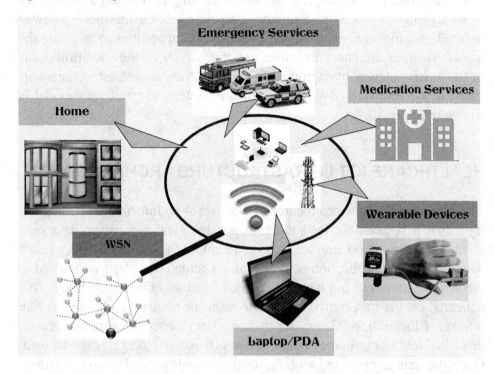

gateway has to provide mechanism to join into the WSN. Through this part, smart gateway provides connection to WSN in both hardware and software ways. At hardware way, smart gateway has to handle interface compatible and signal translation and rate conversion. For software, it has to translate the protocol using in WSN and extract the valuable data which can be used by the program using in smart gateway (Serdaroglu & Baydere, 2015). In simple model, gateway acts only as a protocol translator and data transmitter. Architecture of simple mode is shown in figure 10. It just connects to the WSN on one side and HCC on the other, and transmits data packets between these two sides without storing them. Elderly health status monitoring, data storing, mote status monitoring, and send alarm and question message to hospital and relative works are pass to the central server.

Intelligent model is on the contrary to it. Smart gateway becomes the central part of the whole system. It takes responsible for most of the data processing works, like data analysis and storage, elderly status, mote status and WSN status monitor works are distributed to it. In normal situation in intelligent model, gateway sent health status report to HCC regularly (Allegretti, 2014).

Alarm and question messages will also be sending to hospital and relatives by smart gateway when those situations are detected. Furthermore, gateway in intelligent model also provide WLAN access capabilities to facilitate the use of system maintainers and elder's relatives to check the operation status of the smart gateway by laptop or PDA when they are closed to the smart gateway. Architecture of WSN Healthcare System in intelligent model is shown in figure 10.

HEALTHCARE IOT INFRASTRUCTURE ARCHITECTURE

The components of the healthcare IoT consists of an Information perception layer that is responsible for collecting physiological parametric data such as temperature, blood pressure, oxygen saturation, heartbeat rate, etc. of patients. The data collection is by sensors or actuators that are connected to medical wearables or implantable devices fixed into the human body. This collected data is transmitted via wireless sensor network technologies like infrared, Bluetooth, NFC, etc. to a dedicated device such as a mobile phone, dedicated monitor or home computer that is then used to aggregate the data. Once the data is collected and aggregated (Kashyap & Piersson, 2018b), it is transferred to the network transmission layer that is responsible for carrying the data to the application layer that includes a remote HIS service. Conventional internet technologies such as data networks, wired or wireless connection are used in transmitting the data (Chachin, 2017). The collected information is then processed at the application layer and sent back via the same communication channel to the patient.

We see from the described communication flow that every layer of the architecture has a potential to be exploited because they all serve as entry points into the system. If incorrect data is sent to the application layer and if appropriate data correlation techniques are not implemented and erroneous data gets undetected, then there is a possibility that a patient's life can be in danger. Also, if the functioning of a health information system (server) that returns a processed data to the client is tampered with, the patient still gets exposed to danger. The communication channel that transmits data from the patient to the remote HIS can also be an entry point to intercept the data that is being communicated. An attacker may gain remote or physical access to the HIS to modify the data that is being sent or better servers can be affected by a denial of service attack or even ransom ware attacks which have been a common trend in healthcare IoT industry within the last two years.

Security Issues of Healthcare IoT Infrastructure

Wireless Sensor Network Attacks

The data collection layer in healthcare IoT is exploited such that the data that is sent to the neighboring devices is intercepted and manipulated causing non-integrity, or non-confidentiality of data (Diaz & Sanchez, 2016). This layer consists of RFIDs, sensors, and actuators. Attacks can come in the form of the following:

Sensor Jamming

This can be used to deny communication between two devices. With this, an attacker can prevent a device from communicating with another device and then connect a rogue device in the process in what can be a man in the middle attack as explained. The chapter further describes how a Bluetooth enabled mobile device was jammed so that a rogue device can then communicate with the healthcare IoT implantable device to stage a replay attack.

Eavesdropping

An attacker can intercept communication between devices by sniffing the packets and using the data collected to the advancement of another attack. In the attack illustrated in, there was a need to capture the packets to know the packet size and what and what needs to the manipulated

Spoofing

An attacker can use manipulate a device to think that it is communicating with another authenticated pair by broadcasting a rogue device as the original device. The researchers showed a simulation of an attack on a selected health IoT application by spoofing a device to establish a Bluetooth connection. Masquerading like an original device also makes the attack difficult to detect as the communicating device thinks it is communicating with an authenticated device, an attacker would have used other methods to generate the authentication credentials.

Data Aggregators Vulnerabilities

The data aggregators are classified with sensor-enabled devices as part of the BANs. Mobile smartphones, dedicated health mobile devices, and desktop health monitors etc. all fall into this category and is a threat to the healthcare IoT system. These devices come with dedicated software applications that are pre-configured to match the operation of a sensor device. Misconfiguration can lead to a device being exploited for an attack. The software can also be exposed to malware attacks since these devices are connected to the internet. If physical security is breached, a malicious user can physically send erroneous data through the device also since patients will use these devices, the usability of rigorous security mechanisms may not be convenient.

Social Engineering

Social engineering techniques can be used to gain access physically or remotely into the system. An attacker can disguise as a sick patient to gain access into a hospital and then use rogue means to discover how a hospital IT system is being set-up. He can also use shoulder surfing technique to discover the access credentials of the system. Although this may not be very realistic, a motivated and well-coordinated attack might be successful.

DISCUSSION

This exploit semantic web technologies for several reasons firstly, semantics enables an explicit description of the meaning of sensor data in a structured way, so that machines could understand it. Secondly, it facilitates interoperability for data integration since heterogeneous IoT data is converted according to the same vocabulary. Thirdly, semantic reasoning engines can be easily employed to deduce high-level abstractions from sensor data. Fourthly, context-awareness could be implemented using semantic reasoning. Finally, in theory, semantics eases the knowledge sharing and reuse of domain knowledge expertise which should avoid the reinvention of the wheel. Indeed, each time a new domain specific vocabulary is defined. Semantic web technologies are becoming very popular and are adopted by companies such as Google and Yahoo. Google introduces the idea of the knowledge graph to connect and structure the data with each other. Moreover, 'Linked data' is more and more popular to share and reuse data to build and enhance rich web applications with little effort. The benefits of the ubiquitous connectivity in the Healthcare industry cannot be over-emphasized. In-fact it outweighs the possible attacks predicted in this chapter considering the over 2.5 million people that rely on IMDs in the United States. However, an increasing number of users come with increased attention from manufacturers, security researchers, attackers and defenders. It is therefore important to understand some of the threats that are likely in this domain. The outcome of our research shows that an attack does not require a high sophistication to generate a physical result. Contrary to the claims that cyber terrorism is not a near threat due to sophistication of attacks; this chapter has helped to understand clearly that there is a possibility. Terrorist organizations can opt for the less sophisticated attacks that will be less expensive to carry out and require a moderate level of technical-know how. In the United Kingdom Cyber Security Strategy document for the year 2016 – 2021, it was highlighted that the technical capabilities of terrorists remain limited while they aim to destabilize the computer network operations in the UK, publicity and disruption remains their cyber goal. The same document also emphasizes that for us to measure the success of a government in preventing terrorism, there is a need to fully understand the risk posed by cyber terrorism and the cyber threat from terrorist actors and hostile nation states. This can be achieved through identification and investigation of cyber terrorism threats. This chapter is a valuable input that addresses some of the concerns raised in the cyber security strategy document as it lays emphasis on attacks that create physical results which is one of the goals of a terrorist organization. The output of this chapter further proves the statement that "terrorists will likely use any cyber capability to achieve the maximum effect possible". "Thus, even a moderate increase in terrorist capability may constitute a significant threat to a state and its interest".

CONCLUSION

The rapid development of technology and the Internet leads to growing applications of new technological solutions at the global level. With the appearance of the Internet of Things concept, elements, such as sensors and sensor networks, are becoming available and applicable in all fields of human activity, thus providing conditions for the creation of expert systems that can operate anytime and anywhere. Following these trends, an indispensable application of it is in healthcare where the application can be found in health monitoring, diagnostics and treatment more personalized, timely and convenient. All of this significantly improves health by increasing the availability and quality of care followed with radically reduced costs. Due to the lengthening of life expectancy, society is aging, and more and more people live to an older age. Therefore, it is highly important to assure life quality and safety. Existing and emerging technologies can provide tools that can support elderly people in their everyday life, making it easy and safe. This chapter concerns the design methodology of such tools especially for indoor and outdoor localization, health monitoring, fall detection and behavior recognition and classification. The authors propose the design methodology for the IoT-based home care information system intended for indoor and outdoor environment use. The presented DM approaches the home care problem not only from the designer's perspective, but also considering the contracting authority's and potential users' requirements, which means that apart from the technical requirements, the design procedure considers the multifarious constraints, including the lifetime, energy issue, usage comfort and even the price. The DM was verified with a case study of real-life scenarios where there is a need for supporting elderly people, especially those of limited mobility living alone. The desire stated by the stakeholders and future users required the system for identifying people's position and their vital signs, but also to be able to recognize basic activities, especially falls, and to classify them as normal, suspicious or dangerous. The smart gateway design prototype for health care system using WSN in this smart gateway design, tasks like sensor data storage, elder's current health state detection and real-time report are done in the low power embedded system in the intelligent model. Hardware and software design of the gateway are presented and transmit protocol is designed for this gateway-central system architecture. In this chapter, we focus on the interoperability of sensor data to build promising and interoperable domain-specific or cross-domain IoT applications.

REFERENCES

Abdelwahab, S., Hamdaoui, B., Guizani, M., & Rayes, A. (2014). Enabling Smart Cloud Services Through Remote Sensing: An Internet of Everything Enabler. *IEEE Internet Of Things Journal*, *1*(3), 276–288. doi:10.1109/JIOT.2014.2325071

Allegretti, M. (2014). Concept for Floating and Submersible Wireless Sensor Network for Water Basin Monitoring. *Wireless Sensor Network*, *06*(06), 104–108. doi:10.4236/wsn.2014.66011

Aminzade, M. (2018). Confidentiality, integrity and availability – finding a balanced IT framework. *Network Security*, *2018*(5), 9–11. doi:10.1016/S1353-4858(18)30043-6

Brinis, N., & Saidane, L. (2016). Context Aware Wireless Sensor Network Suitable for Precision Agriculture. *Wireless Sensor Network*, *8*(1), 1–12. doi:10.4236/wsn.2016.81001

Chachin, P. (2017). IoT is being introduced into housing and utilities infrastructure. *Electronics: Science, Technology, Business*, (6), 138–142. doi:10.22184/1992-4178.2017.166.6.138.142

Chao, C., & Hsiao, T. (2014). Design of structure-free and energy-balanced data aggregation in wireless sensor networks. *Journal of Network and Computer Applications*, *37*, 229–239. doi:10.1016/j.jnca.2013.02.013

Chen, C. (2016). A Fuzzy Indoor Positioning System with ZigBee Wireless Sensors. *Journal of Electrical and Electronics Engineering (Oradea)*, *4*(5), 97. doi:10.11648/j.jeee.20160405.12

Condon, S. (2018). IoT will account for nearly half of connected devices by 2020, Cisco says. *ZDNet*. Retrieved from http://www.zdnet.com/article/iot-will-account-for-nearly-half-of-connected-devices-by-2020-cisco-says

Diaz, A., & Sanchez, P. (2016). Simulation of Attacks for Security in Wireless Sensor Network. *Sensors (Basel)*, *16*(11), 1932. doi:10.339016111932 PMID:27869710

Dogan, G. (2016). ProTru: A provenance-based trust architecture for wireless sensor networks. *International Journal of Network Management*, *26*(2), 131–151. doi:10.1002/nem.1925

Domingo, M. (2014). Sensor and gateway location optimization in body sensor networks. *Wireless Networks*, *20*(8), 2337–2347. doi:10.100711276-014-0745-7

Ghoneim, M., & Hussain, M. (2015). Review on Physically Flexible Nonvolatile Memory for Internet of Everything Electronics. *Electronics (Basel)*, *4*(3), 424–479. doi:10.3390/electronics4030424

Hejlová, V., & Voženílek, V. (2013). Wireless Sensor Network Components for Air Pollution Monitoring in the Urban Environment: Criteria and Analysis for Their Selection. *Wireless Sensor Network*, *5*(12), 229–240. doi:10.4236/wsn.2013.512027

Kashyap, R., & Gautam, P. (2017). Fast Medical Image Segmentation Using Energy-Based Method. *Biometrics. Concepts, Methodologies, Tools, and Applications*, *3*(1), 1017–1042. doi:10.4018/978-1-5225-0983-7.ch040

Kashyap, R., Gautam, P., & Tiwari, V. (2018). Management and Monitoring Patterns and Future Scope. In Handbook of Research on Pattern Engineering System Development for Big Data Analytics. IGI Global. doi:10.4018/978-1-5225-3870-7.ch014

Kashyap, R., & Piersson, A. (2018a). Big Data Challenges and Solutions in the Medical Industries. In Handbook of Research on Pattern Engineering System Development for Big Data Analytics. IGI Global. doi:10.4018/978-1-5225-3870-7.ch001

Kashyap, R., & Piersson, A. (2018b). *Impact of Big Data on Security. In Handbook of Research on Network Forensics and Analysis Techniques* (pp. 283–299). IGI Global. doi:10.4018/978-1-5225-4100-4.ch015

Kashyap, R., & Tiwari, V. (2018). Active contours using global models for medical image segmentation. *International Journal of Computational Systems Engineering*, *4*(2/3), 195. doi:10.1504/IJCSYSE.2018.091404

Kuang, J., Niu, X., & Chen, X. (2018). Robust Pedestrian Dead Reckoning Based on MEMS-IMU for Smartphones. *Sensors (Basel)*, *18*(5), 1391. doi:10.339018051391 PMID:29724003

Li, J., Jin, J., Yuan, D., & Zhang, H. (2018). Virtual Fog: A Virtualization Enabled Fog Computing Framework for Internet of Things. *IEEE Internet Of Things Journal*, *5*(1), 121–131. doi:10.1109/JIOT.2017.2774286

Monika, & Upadhyaya, S. (2015). Secure Communication Using DNA Cryptography with Secure Socket Layer (SSL) Protocol in Wireless Sensor Networks. *Procedia Computer Science, 70*, 808-813. doi:10.1016/j.procs.2015.10.121

Morain, S., Kass, N., & Grossmann, C. (2016). What allows a health care system to become a learning health care system: Results from interviews with health system leaders. *Learning Health Systems, 1*(1), e10015. doi:10.1002/lrh2.10015

Papers, W. (2018). *Cisco Visual Networking Index: Forecast and Methodology, 2016–2021.* Retrieved from http://www.cisco.com/c/en/us/solutions/collateral/service-provider/visual-networking-index-vni/complete-white-paper-c11-481360.html

Peng, S. (2015). Cloud-Based Sport Training Platform Based on Body Sensor Network. *Journal Of Software Engineering*, *9*(3), 586–597. doi:10.3923/jse.2015.586.597

Pesko, M., Smolnikar, M., Vučnik, M., Javornik, T., Pejanović-Djurišić, M., & Mohorčič, M. (2014). Smartphone with Augmented Gateway Functionality as Opportunistic WSN Gateway Device. *Wireless Personal Communications*, *78*(3), 1811–1826. doi:10.100711277-014-1908-7

Scott, R., & Mars, M. (2013). Principles and Framework for eHealth Strategy Development. *Journal of Medical Internet Research*, *15*(7), e155. doi:10.2196/jmir.2250 PMID:23900066

Serdaroglu, K., & Baydere, S. (2015). WiSEGATE: Wireless Sensor Network Gateway framework for internet of things. *Wireless Networks*, *22*(5), 1475–1491. doi:10.100711276-015-1046-5

Snyder, T., & Byrd, G. (2017). The Internet of Everything. *Computer*, *50*(6), 8–9. doi:10.1109/MC.2017.179

Wang, X., & Zhang, Z. (2015). Data Division Scheme Based on Homomorphic Encryption in WSNs for Health Care. *Journal of Medical Systems*, *39*(12), 188. doi:10.100710916-015-0340-1 PMID:26490146

Watanabe, K., Fukuda, K., & Nishimura, T. (2015). A Technology-Assisted Design Methodology for Employee-Driven Innovation in Services. *Technology Innovation Management Review*, *5*(2), 6–14. doi:10.22215/timreview/869

Xu, L., Collier, R., & O'Hare, G. (2017). A Survey of Clustering Techniques in WSNs and Consideration of the Challenges of Applying Such to 5G IoT Scenarios. *IEEE Internet Of Things Journal*, *4*(5), 1229–1249. doi:10.1109/JIOT.2017.2726014

Zhang, P., Nagarajan, S., & Nevat, I. (2017). Secure Location of Things (SLOT): Mitigating Localization Spoofing Attacks in the Internet of Things. *IEEE Internet Of Things Journal*, *4*(6), 2199–2206. doi:10.1109/JIOT.2017.2753579

KEY TERMS AND DEFINITIONS

IHH: In home healthcare service. Human administrations at home is a pioneer in bringing modified and capable home therapeutic administrations benefits in India to allow quick and accommodating recovery inside the comfort of one's home. A part of the key helpful organizations offered consolidate setting up ICU at home, giving cancer care at home, nursing care, physiotherapy organizations and widely inclusive stroke recuperation close by giving a lot of clinical system at home along these lines passing on pretty much 70% of each clinical organization at home.

IoT: The internet of things (IoT) is the system of physical gadgets, vehicles, home apparatuses, and different things installed with hardware, programming, sensors, actuators, and availability which empowers these things to associate and trade information, making open doors for more straightforward coordination of the physical world into pc based frameworks, bringing about proficiency upgrades, financial advantages and lessened human intercession.

Chapter 3
Strengthening Agriculture Through Energy–Efficient Routing in Wireless Sensor Networks Using Sink Mobility

Subba Reddy Chavva
VIT-AP University, India

Nagesh Mallaiah Vaggu
VIT-AP University, India

Ravi Sankar Sangam
VIT-AP University, India

ABSTRACT

Wireless sensor networks (WSNs) can be used in agriculture to provide farmers with help monitoring the fields. Most of the people depend on agriculture. WSN plays a vital role in strengthening agriculture. In this chapter, the authors discuss energy-efficient routing with mobile sink protocols that are more suitable to strengthen the agriculture. They organize this chapter by classifying aforesaid protocols into three different categories (e.g., hierarchical-based, tree-based, and virtual structure-based routing).

DOI: 10.4018/978-1-5225-9004-0.ch003

INTRODUCTION

Agriculture exists since past many centuries and that has played vital role in human evolution. The quality of sedentary human civilization was greatly enhanced by farmers in farming agriculture. Since the nation's economic growth depends on Agriculture, there is an enormous need for improving crop yield production with innovative technologies (Mendez, Yunus & Mukhopadhyay, 2012). The technologies that support the farming procedures to enhance the ease of farming are surpassing day by day. Wireless Sensor Networks (WSNs) as an emerging technology can be used in farming to further enhance the ease of farming. WSNs can be used in farming to monitor environmental conditions such as humidity, temperature, moisture levels, atmospheric pressure, and soil water to maintain the health of plants. These nature parameters are necessary in growing, strengthening of plants. Besides, WSNs also useful in sensing the early disease of plants that will leap away from disasters. This technology will facilitate farmers to do farming with less time and efforts with more profits (Al, Braik & Bani-Ahmad, 2010; Deepika & Rajapirian, 2016).

In WSNs, all sensor nodes have limited energy to send data to gateways, also sometimes referred as sink nodes, and receive acknowledgments from sink nodes. So, decreasing energy consumption and increasing throughout of a network is needed. When sink node or base station is initiated to all other sensor nodes with a request for data. All the sensor nodes send data directly to sink node, it consumes a lot of energy. So, effective routing technique is required among sensor nodes and sink node. An enormous amount of sensor data evolved from the terrestrial nodes often have no longer used data from base station with exponentially creating leverage energy consumption. Effective routing is important for energy efficiency in network. Routing designs a path for sensor nodes, sends data to sink node or base station or gateway. To decrease energy consumption of sensor nodes, dynamic change of network routing is required (Vu, Nguyen & Nguyen, 2014). The overview of Wireless Sensor Network (WSN) is shown in Figure 1. In a WSN some numbers of wireless sensor nodes are deployed. Among these sensor nodes some of them are active and transmit data to the base station, through Internet or Wi-Fi and furthermore the base station forwards these data to user. The nodes that are with below threshold energy are called as exhausted sensor nodes. The standby nodes are the nodes with full energy and not involved in communication.

Figure 1. Overview of wireless sensor network (WSN)

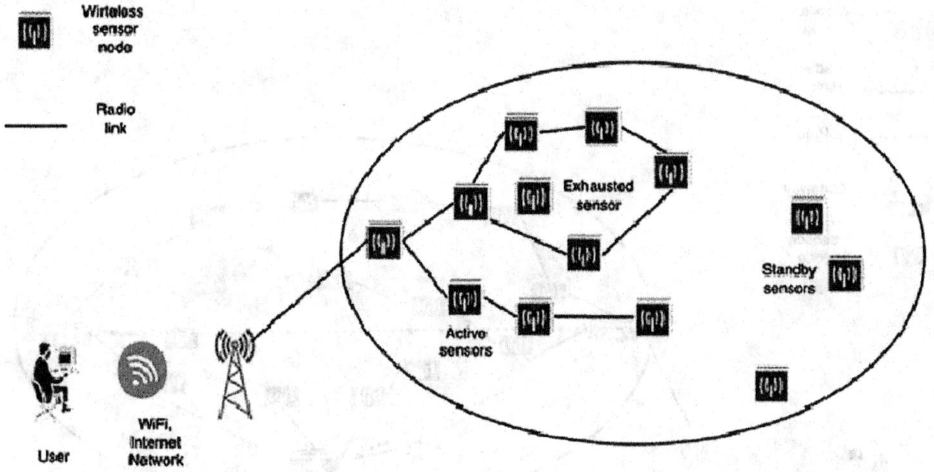

ROUTING IN WIRELESS SENSOR NETWORK

Routing performs significant task in WSN. In WSN, it is not very trivial to assign global ids due to large number of sensor nodes in network. Although there exist many conventional routing protocols but they are less efficient to prolong the network life time, as sensor nodes consists less battery power, storage and processing capacity. In other words, the amount of data that a node transfers is directly propositional to consumption of energy (Xie, Shi, Hou, Lou, Sherali & Midkiff, 2015; Jain, Saini & Bhooshan, 2015).

One major challenge in developing energy-efficient routing protocols is that the sensor nodes near to sink die fast due to data transfer overhead. Here, it may be noted that some sensor nodes in the network still have some energy but they cannot be operative. This phenomenon is called as *"Hotspot Problem"* or *"Crowded Center Effect"* (Popa, Rostamizadeh, Karp, Papadimitriou & Stoica, 2007) or *"Energy Hole Problem"* (Li & Mohapatra, 2007). Therefore, it is rattling significant to have energy-efficient routing protocols to effectively transfer sensed data from nodes to base station in WSN without aforesaid problem.

The concept of sink mobility is very useful to efficiently address the hotspot problem. In sink mobility the routing process is dynamic as the sink node changes their positions at regular time intervals (Li & Mohapatra, 2007). The process of routing in WSN using sink mobility is depicted in Figure 2. In the process of sink mobility, all the sensor nodes form different route paths to

Figure 2. Process of routing in WSN using sink mobility

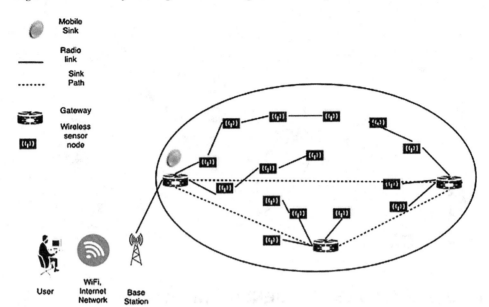

different gateway nodes. The sink node moves with in the network around gateways and it collect and transmit data to the base station. Now, base station forwards these data to the user through Internet or Wi-Fi.

In this chapter, we are aimed to discuss three different dynamic protocol classifications viz. hierarchical based, tree based and virtual-structure based routings. These sink mobility routing protocols are most suitable for strengthening of agriculture.

HIERARCHICAL BASED ROUTING

Network in hierarchical routing is organized as levels. Even though the layers are close to the target data source the superior layer will take the responsibility of processing and transmitting sensed data from the source. The lower layer will transmit the information to the higher layer yielding the queries that will go from top to bottom. In the hierarchical approach, imaginary hierarchical nodes are produced that will employ distinct robust function on the sensors. A level of the hierarchy is structured as two or more. An effective hierarchical towards energy hole problem (Li & Mohapatra, 2007) will be suppressed by keeping off it on the top level nodes.

The cluster-based mobile routing architecture has been proposed in (Wang, Huang, Fu & Wang, 2008). In this architecture, the neighbor nodes are identified by sending a control packet from each sensor node. Based on the remaining residual energy of nodes, the node with highest remaining residual energy will act as cluster head for its neighbor nodes. To form a cluster, the cluster head advertises packets to the neighbors. The accepted neighbors will form a cluster. In this protocol, the random waypoint mobility model has been employed for the movement of the mobile sink. The information about the location is sent by the sink node to the cluster head. Based on this location address the cluster head transmit the collected data using laid routing path.

Energy-aware data aggregation scheme was proposed in (Wang, Yeh & Huang, 2007). Authors have proposed routing protocol with hierarchical hybrid method that contains tree data with a shape of grids. Sensor nodes are enabling with Global Positioning System (GPS) that facilitates the nodes to determine the nearest sink (Cheng, Du, Yang, Yu, Chen & Guan, 2011). The nearest sink with energy greater than all other nodes energy will be elected as a Gateway node. The sensed data received by a Gateway node will aggregate data, to lower the utilization of energy, and send to nearest sink node in the routing path. It may be noted that the routing path may change at regular time intervals according to mobility of the sink. In this protocol the velocity of a sink node will be maximized to minimize the energy consumption.

TREE BASED ROUTING

Tree based routing management has tremendously improved the performance in sink mobility than hierarchical based routing. Through the joined structure similar to a tree, it effectively manages the mobility of sink. Whatever the source node transmits information to sink node the tree structure will transmit on nominal rate.

In (Kim, Abdelzaher & Kwon, 2003), a Scalable energy-efficient asynchronous dissemination protocol is proposed in which tree structured routing path has been used to pass the data across access nodes to sink node. To transmit data along the path the access nodes should grab the address of next node in that path. The dynamic mobility is achieved by continuous monitoring of the preceding nodes called as new access nodes its address. Restructuring of the tree is updated whenever the sink node is attached with new position (new access node location). Unbalancing will be faced by such protocol within delay reduction and crediting strength drop on tree reconfiguration.

In this case mobility of sink node is achieved by avoiding the update position of node. This concept is well applicable in real-time applications. Entire cost will be enhanced by the sink that opts its modern access node.

Sharma & Jena (2014) have proposed a tree based Data Dissemination protocol with Mobile Sink (TEDD) to overcome higher routing overhead and shorter lifetime. TEDD consist of two categories of nodes in the tree: Relay Node (RN), Non-Relay Node (Non-RN). Relay node is responsible to send data to next relay node and non-relay node is responsible to send data to relay node. Non-relay nodes are unidirectional communicators and relay nodes are bi-directional communicators. In this protocol, sensor nodes data is send to sink node via a tree structured path. In this tree structure, non-leaf node is called relay node and leaf node is called non-relay node. TEED manages the mobility of sink and balances the load among sensor nodes to maximize the lifetime. TEDD consists of three phases i) neighbor discovery, ii) tree construction and relay node selection, iii) data transformation. In neighbor discovery phase, each sensor node finds the neighbor node and maintains neighbor node list. Sensor node sends request, to act as relay node, to neighbor node. If neighbor node accepts this request, then the relay node list will be updated, by base station, based on residual energy of sensor nodes. In the second phase, based on the neighbor discovery list, tree construction is initiated by initiator node (a nearest node to the base station) by sending the tree message control packet to its neighbor node with highest residual energy, this node will be acting as a relay node. This relay node further sends tree message control packet to its neighbor nodes. In this process, if at least any one of the relay nodes dies then initiator node again initiates tree construction by sending tree control message. Each non-relay makes reverse link to its neighbor relay node for sending data. The mobile sink moves within the network and collects data through gateways. The sink chooses nearest relay or non-relay node as a gateway. The mobile sink moves with in the random waypoint mobility model. After tree construction and relay node selection, data sent via relay node to next relay node and non-relay node to relay node. Eventually, relay nodes transmit data to gateway and gateway forwards it sends to sink to base station via sink node. This protocol can efficiently manage the sink mobility.

A distributed dynamic shared tree protocol was proposed in (Hwang, In & Eom, 2006). This protocol facilitates fast data transfer by using multi sink access. The root node of the tree is called master sink node and by using new location of other slave sink nodes the tree is build. In the network, many slave sink nodes are controlled by master sink node. The data is transmitted from sensor node to master sink node through slave sink nodes.

A Multi-Point Relay (MPR) based routing protocol called SN-MPR has been proposed in (Faheem & Boudjit, 2010; Qayyum, Viennot & Laouiti, 2002). Large numbers of sensor nodes are deployed in a network and the network is divided into two types: non-MPR and MPR nodes. Based on the residual energy, it chooses the MPR node for data transmission, when sink want to collect data from sensor nodes. Neighbor node receives location of update by the sink broadcast. Only MPR nodes transfer data to sink nodes. MPR nodes get the updated sink position, by a broadcast message from sink, in timely manner whenever sink node(s) changes their position and accordingly they create reverse link towards sink node. Hence, energy consumption is very high; because of the tree structure is affected whenever movement of the sink takes place.

VIRTUAL-STRUCTURE BASED ROUTING

A virtual infrastructure is often considered as an efficient approach for data dissemination over the network with mobile sinks. The concept of virtual infrastructure acts as a rendezvous area for storing and retrieving the collected data (Luo, Ye, Cheng, Lu & Zhang, 2005). The sensor nodes that are belonging to rendezvous area are responsible for storing the sensory data during the absence of sink nodes. As soon as mobile sink moves from one grid to another grid the selected nodes belonging to that rendezvous area transfer stored sense data to mobile sink. This virtual infrastructure can be built using a backbone based or a rendezvous-based approach.

In (Luo, Ye, Cheng, Lu & Zhang, 2005), a two-tier data dissemination protocol, an energy efficient protocol for large scale WSN was proposed. In this protocol, huge number of sensor nodes forms a grid infrastructure and also it supports multiple sinks. Sensor nodes are intended to send data to the sink node and the protocol creates a virtual grid. The virtual grid has some crossing points, and these points are selected by the dissemination nodes. These nodes are very useful when an event occurs to transmits data by using source id. The mobile sink request to procure information and these queries goes through virtual grids. The sink node receives data from source node by using reverse path. However, there is an overhead on energy consumption for constructing multiple routing paths from sensor nodes to mobile sink in a grid consumes.

Clustered Tree based Routing protocol with Mobile Sink (CTRP) proposed in (Bagga, Sharma, Jain & Sahoo, 2015) to handle WSN with large number of sensor nodes. It divides the network into number of equal size virtual grids. All the sensors in each grid act as cluster members (children nodes) of that grid and one node with highest residual energy in grid/cluster will chose as cluster head. A tree like routing path is constructed among cluster heads. It may be noted that initiator cluster head will initiate this routing path construction. Communication between parent nodes and children nodes are bidirectional. The sink collecting data from each cluster head in the way of random mobility model. The cluster heads are responsible for collecting data from their corresponding cluster members, aggregating collected data to remove redundancy and transferring aggregated data to mobile sink. In any grid if cluster head residual energy goes below a specified threshold energy, again cluster head selection in that grid will be initiated to balance the load among cluster heads. It will decrease energy consumption and increases the lifetime of the network.

Rendezvous based Routing protocol with mobile sink is discussed in (Sharma, Puthal, Jena, Zomaya & Ranjan, 2017). There is large number of sensor nodes deployed in a network. Each sensor node finds its neighbor nodes and places its neighbor table list in sensor nodes. The large network is divided into equal parts called vertical and horizontal strips. The strips are divided by some area known as cross area. The nodes that are present in this area called back bone-tree nodes. Based on the neighbor list of sensor nodes and their residual energy, a back bone-tree node forms a routing path in the rendezvous region. The tree construction starts from centroid of the rendezvous region. The centroid node selects closest neighbor node based on the neighbor list and from tree up to boundary nodes. There are two types of data transmission methods for data transfer from sensor node to sink node. In first method, sensor nodes send data to sink via identified nearer backbone-tree nodes. In second method, sensor nodes send data, based on sink position, to sink via backbone-tree nodes that are closer to sink. The sink always moves in the network with the random waypoint mobility model. When it reaches a new location then the sink broadcasts its position to back bone-tree nodes.

CONCLUSION

On reviewing the major routing protocols inclusive of sink mobility in wireless sensor networks for strengthening the agriculture, these techniques ensure energy efficiency within the nodes along with reducing control packet overhead. Based on the foregoing discussion, we conclude that packet congestion is minimized along with energy optimization among sensor nodes. Hence dynamic rearrangement of routing tree makes the system more efficient towards network delay and improving of network lifetime.

REFERENCES

Al Bashish, D., Braik, M., & Bani-Ahmad, S. (2010, December). A framework for detection and classification of plant leaf and stem diseases. *IEEE International Conference on Signal and Image Processing (ICSIP)*, 113-118. doi:10.1109/ICSIP.2010.5697452

Bagga, N., Sharma, S., Jain, S., & Sahoo, T. R. (2015). A cluster-tree based data dissemination routing protocol. *Procedia Computer Science, 54*, 7–13. doi:10.1016/j.procs.2015.06.001

Cheng, B., Du, R., Yang, B., Yu, W., Chen, C., & Guan, X. (2011, September). An accurate GPS-based localization in wireless sensor networks: a GM-WLS method. *40th IEEE International Conference on Parallel Processing Workshops (ICPPW)*, 33-41. doi:10.1109/ICPPW.2011.32

Deepika, G., & Rajapirian, P. (2016, February). Wireless sensor network in precision agriculture: a survey. *IEEE International Conference on Emerging Trends in Engineering, Technology and Science (ICETETS)*, 1-4. doi: 10.1109/ICETETS.2016.7603070

Faheem, Y., & Boudjit, S. (2010, December). SN-MPR: A multi-point relay based routing protocol for wireless sensor networks. *IEEE/ACM Int'l Conference on Green Computing and Communications (GreenCom) & Int'l Conference on Cyber, Physical and Social Computing (CPSCom)*, 761-767. doi:10.1109/GreenCom-CPSCom.2010.139

Hwang, K. I., In, J., & Eom, D. S. (2006, February). Distributed dynamic shared tree for minimum energy data aggregation of multiple mobile sinks in wireless sensor networks. In *European Workshop on Wireless Sensor Networks* (pp. 132-147). Springer. doi:10.1007/11669463_12

Jain, T. K., Saini, D. S., & Bhooshan, S. V. (2015). Lifetime optimization of a multiple sink wireless sensor network through energy balancing. *Journal of Sensors*.

Kim, H. S., Abdelzaher, T. F., & Kwon, W. H. (2003, November). Minimum-energy asynchronous dissemination to mobile sinks in wireless sensor networks. *Proceedings of the 1st ACM international conference on Embedded networked sensor systems*, 193-204. doi:10.1145/958491.958515

Li, J., & Mohapatra, P. (2007). Analytical modeling and mitigation techniques for the energy hole problem in sensor networks. *Pervasive and Mobile Computing*, *3*(3), 233–254. doi:10.1016/j.pmcj.2006.11.001

Luo, H., Ye, F., Cheng, J., Lu, S., & Zhang, L. (2005). TTDD: Two-tier data dissemination in large-scale wireless sensor networks. *Wireless Networks*, *11*(1-2), 161–175. doi:10.100711276-004-4753-x

Mendez, G. R., Yunus, M. A. M., & Mukhopadhyay, S. C. (2012, May). A WiFi based smart wireless sensor network for monitoring an agricultural environment. *IEEE International Instrumentation and Measurement Technology Conference (I2MTC)*, 2640-2645. doi:10.1109/I2MTC.2012.6229653

Popa, L., Rostamizadeh, A., Karp, R., Papadimitriou, C., & Stoica, I. (2007, September). Balancing traffic load in wireless networks with curveball routing. *Proceedings of the 8th ACM international symposium on Mobile ad hoc networking and computing*, 170-179. doi:10.1145/1288107.1288131

Qayyum, A., Viennot, L., & Laouiti, A. (2002, January). Multipoint relaying for flooding broadcast messages in mobile wireless networks. *Proceedings of the 35th Annual Hawaii IEEE International Conference on System Sciences (HICSS)*, 3866-3875. doi:10.1109/HICSS.2002.994521

Sharma, S., & Jena, S. K. (2014). Data dissemination protocol for mobile sink in wireless sensor networks. *Journal of Computational Engineering*, *2014*, 1–10. doi:10.1155/2014/560675

Sharma, S., Puthal, D., Jena, S. K., Zomaya, A. Y., & Ranjan, R. (2017). Rendezvous based routing protocol for wireless sensor networks with mobile sink. *The Journal of Supercomputing*, *73*(3), 1168–1188. doi:10.100711227-016-1801-0

Vu, T. T., Nguyen, V. D., & Nguyen, H. M. (2014). An energy-aware routing protocol for wireless sensor networks based on k-means clustering. In *Recent Advances in Electrical Engineering and Related Sciences (AETA)* (pp. 297–306). Berlin: Springer. doi:10.1007/978-3-642-41968-3_31

Wang, N. C., Yeh, P. C., & Huang, Y. F. (2007, August). An energy-aware data aggregation scheme for grid-based wireless sensor networks. *Proceedings of the ACM international conference on Wireless communications and mobile computing*, 487-492. doi:10.1145/1280940.1281044

Wang, Y. H., Huang, K. F., Fu, P. F., & Wang, J. X. (2008, June). Mobile sink routing protocol with registering in cluster-based wireless sensor networks. In *International Conference on Ubiquitous Intelligence and Computing* (pp. 352-362). Springer. doi:10.1007/978-3-540-69293-5_28

Xie, L., Shi, Y., Hou, Y. T., Lou, W., Sherali, H. D., & Midkiff, S. F. (2015). Multi-node wireless energy charging in sensor networks. *IEEE/ACM Transactions on Networking*, *23*(2), 437–450. doi:10.1109/TNET.2014.2303979

Chapter 4

A Novel Power–Monitoring Strategy for Localization in Wireless Sensor Networks Using Antithetic Sampling Method

Vasim Babu M.

KKR and KSR Institute of Technology and Sciences, India

ABSTRACT

The prime objective of this chapter is to develop a power-mapping localization algorithm based on Monte Carlo method using a discrete antithetic approach called Antithetic Markov Chain Monte Carlo (AMCMC). The chapter is focused on solving two major problems in WSN, thereby increasing the accuracy of the localization algorithm and discrete power control. Consecutively, the work is focused to reduce the computational time, while finding the location of the sensor. The model achieves the power controlling strategy using discrete power levels (CC2420 radio chip) which allocate the power, based on the event of each sensor node. By utilizing this discrete power mapping method, all the high-level parameters are considered for WSN. To improve the overall accuracy, the antithetic sampling is used to reduce the number of unwanted sampling, while identifying the sensor location in each transition state. At the final point, the accuracy is increased to 94% wherein nearly 24% of accuracy is increased compared to other MCL-based localization schemes.

DOI: 10.4018/978-1-5225-9004-0.ch004

INTRODUCTION

Wireless Sensor Network (WSN) is named as a group of wireless networked low-power sensor devices in which, each node incorporates with microprocessor, radio and a limited amount of storage. The couple of tasks like localizing and tracking, moving stimuli or objects are essential capabilities of a sensor network. The major problems, that are considered in designing the proposed localization algorithm, are high power consumption, cost and time synchronization. Also, localization error, beacon density, 2D analysis structure and low sampling efficiency hinder the performance of the localization algorithms. The existing Adaptive Monte Carlo Technique experiences the foresaid problems. Based on the problem specification the objective of the proposed system is framed to solve the grievances and to obtain the desired optimal results in Wireless Sensor Network. The localization schemes for WSN have been developed in the last 20 years. The schemes have been widely used in various applications like military, civil, multi-robot search teams, automated guided vehicles and many others. These applications consist of multiple autonomous agents locally interacting in pursuit of a global goal. To control over the above system, the distributed control strategy is incorporated. Moreover, the transmission power plays a key role in the design of wireless networks. Power control helps in various functionalities in wireless sensor network which has been stated by Jaein jeong et al. (2007). They are:

- *Interface Management:* In broadcast wireless network, the signals interfere with each other. It is very crucial in CDMA systems where orthogonality between the users is difficult to maintain. In this system, the power control strategy helps the user in efficient spectral reuse and desirable communication experience.
- *Energy management:* The lifetime of the nodes and the network rely on the energy conservation, due to inadequate battery power in mobile stations, hand-held devices and or in any nodes that generally operate on limited energy budget. The energy conservation is made possible through power control strategy.
- *Connectivity management:* In wireless network, the signals are uncertain, energy limited and time-variated. In order to estimate the channel state and to be stay- connected with the transmitter, the receiver should be able to maintain a least possible level of received

signal, due to the above characteristics of wireless network. Here, the power control strategy assists in maintaining a logical connectivity for a given signal processing of a network.

Thus, the power control strategy plays an essential task in WSN. Moreover, the power utilization is varied, due to the location of the sensors. Several researchers have pointed that the efficient performance of WSN is often influenced by localization and power mapping. Localization is crucial and complex for some applications in WSN. Monte Carlo Localization is widely used for localization process and variance reduction techniques like antithetic variates, control variate and conditional variates are used as discussed by James Neufeld et al. (2015), but, these algorithms are not feasible in terms of accuracy and power management. Hence, methodologies for improving the localization process and discrete power mapping including the aspect of variance reduction are highly focused in this paper.

In WSN, the localization process is a crucial task that reflects on the performance efficiency. Mostly, these algorithms are broadly classified into Centralized and Distributed. Generally, the distributed algorithms are more robust and energy efficient than centralized algorithm as stated by Shang et al. (2004). Based on the application, some algorithms assume simple connectivity information between neighbouring nodes. It was employed by Zhou et al. (2004). On the other hand, some applications gather ranging information and angle information. That was described by Shigeng Zhang et al. (2010). The ranging information is defined as the estimated distance between two neighbouring nodes and the angle distance is termed as relative angle distance between received radio signals. The actual or absolute position of each sensor node is determined through a small fraction of special nodes called seeds or anchor nodes with known positions.

Certain techniques have been proposed to determine the position of the nodes but often these methods are influenced by the parameters like accuracy, power consumption, cost, time complexity, memory and many others. Sometimes, the applications of the localization method also reduce the performance.

The centralized localization algorithm usually runs on a base station and all the other nodes in the sensor network send their measured data to the base station. There are certain leading benefits of the centralized algorithm. Some of which are i) It can be designed with more accuracy ii) It can process large amount of data. Further, the algorithm has some common pitfalls like poor scalability, data routing complexity and greater power consumption. More over the distributed algorithms are not fast enough and consume more memory.

According to Heet al. (2003), in Range-based algorithm, a special hardware is needed to accomplish higher localization accuracy and absolute measurement. This hardware is costlier and this limits the range-based algorithm. Although, the Range-free algorithm does not require any special hardware and low cost, the algorithm lacks in performance and accuracy as discussed by singh et al (2017)

Wang & Zhu (2009) have formalized a Sequential Monte Carlo method that uses a non-linear discrete time model for estimation. The SMC based localization algorithm suffers from low sampling efficiency and also requires high beacon density.

Kai Cai et al. (2010) have designed a distributed control of distributed event system for supervisor localization scheme. In this method, global supervisor has been computed and decomposed in to local controllers for investigating the problem of synthesizing the local controller with supervisor node. When the number of supervisor nodes is increased, the synthesis problem may raise as said by Halder and Ghosal (2016)

MCL based localization algorithm is discussed by Guodong Teng et al. (2011). The method has two steps, Prediction and filtering. The former identifies the location in 2D manner with the prior knowledge of sampling states. The latter eliminates the inconsistent data from the current sensor nodes. In this method, the sampling efficiency is not considered due to the reason that, during the prediction time, number of unwanted sensor information is collected and sent to filtering stage.

In some distributed based localization algorithms, the location has been analyzed by some predefined distance measurements and adjusted in to original normalized result of any node whose position is known. It is proposed by Rao Wenbi et al. (2009). The accuracy of this algorithm is better but the computational cost is high. The probabilistic approach has been developed by Weidong Wang et al. (2009) for large scale sensor network to overcome some limitations in single range based technique. The authors have achieved good localization accuracy for large number of sensor nodes. But, the computational time is high because huge numbers of sensor nodes are used as discussed by Abu Znaid et al (2017).

Naraghi Pour et al. (2013) have developed DRGD (Distributed Randomized Gradient Descent) method for sensor network. The location measurement has been taken in noise free environment and provided true location of the nodes with or without distance error. Hence, RSSI parameters can't be considered, during the noise free measurements. To improve the localization accuracy, multiple frequencies and power are used by Xiuyuan Zheng et al. (2014).

Pravena et al. (2014) proposed an Improved Monte Carlo Localization (IMCL) scheme by using bounding box and weight computation technique. In this model, localization failure has been solved by time series forecasting and dynamic sampling method. Due to dynamic sampling method, some unknown node information can't be reached within the localization area at right time and will take large number of samples, when RSS increase. So, the accuracy is little bit low, when assigning dynamic sampling method. But in antithetic sampling along with discrete power control strategy, unnecessary sampling has been totally avoided during the localization period and reached the location at right time with small number of samples.

The energy consumption is also considered to minimize the cost and power terminal levels. Dasgupta et al. (2003), and Barberis et al. (2007) have use the sensor node which uses CC2420 radio chip. The chip controls the irregularities in the signal. To reduce the power control overhead in ad hoc network, DPC (Discrete Power Control) is considered. DPC needs its own problem formulation and analysis. It minimizes the signal-to-noise ratio constraints as stated by Wu & Bertsekas (2001).

The iterative Monte Carlo localization scheme has been developed by Sahoo et al. (2014) for mobile wireless sensor network to improve the positioning accuracy. In this method, every location information will formulate to its neighbour node only once and merge it to a packet then forward the same. The entire information will be lost once the localization error occurs in any state during packet forwarding or while missing some information

Dynamic node model and discrete H_∞ filter based localization algorithm have been developed by Xiaoming Wu et al. (2015) to achieve high reliability and better accuracy. The H_∞ filter uses integration of position information obtained from neighbouring nodes to calculate the position of the nodes. But, the average localization error occurs, while the speed is increased.

Mostly, Monte Carlo Localization (MCL) is used for positioning or locating the sensors as state by Niu, Huan et al (2016). MCL is a recursive Bayes filter which measures the posterior distribution of node's positions. The MCL algorithm possesses two stages. Initially, in the prediction step, the sensor node uses a motion model in order to predict its possible location within a 2D Cartesian space based on the previous samples and their movement. Later, in the filtering step, the node uses a filtering mechanism in order to eliminate those predicted locations that are inconsistent to the current sensor data. It has been discussed by Kurecka et al. (2014), and Anas Alhashimi et al. (2014). The MCL based localization algorithm generates lower estimation error compared to both Centroid and Amorphous localization techniques.

Although MCL algorithm provides higher efficiency in localization process, due to variate signals the performance is hindered and on the other side, the power management is also essential to provide better results as discussed by Anup Kumar Paul (2017). To solve, the Antithetic Monte Carlo Technique is proposed with Discrete Antithetic sampling.

Bennani et al. (2012) implemented the PSO (Particle Swarm Optimization) method to accelerate the convergence speed and to save the energy consumption. Singh et al. (2011) incorporated the Genetic ant colony algorithm which used crossover and mutation operators. The operators avoided the convergence problem in the communication network.

In this paper, a novel technique of antithetic variates is used to reduce variance by presenting negative dependence among the pairs of repetition. This difficulty of self-localization in the sensor networks is renowned as a key constraint for numerous network applications and it can be measured as a significant step in the overall objective of emerging self-adapting and self-configuring networks.

The Antithetic Markov Chain Monte Carlo algorithm (AMCMC) for WSN is based on Antithetic Markov process. This method includes variance reduction technique for improving the precision of Markov Chain Monte Carlo algorithm and also to evaluate a Probability Distribution Function (PDF) with the aid of antithetic Markov mobility model. The ultimate aim of this function is to estimate unknown location of sensor nodes.

PROPOSED LOCALIZATION ALGORITHM USING ANTITHETIC MONTE CARLO METHOD

Localization is termed as locating the geometrical position of sensor node and co-ordination of nodes with one another. It means the accuracy of localization algorithm is extremely essential and the efficiency of the algorithm also depends on the lifetime of the nodes. Here, the antithetic variance reduction technique as discussed in Kroese et al. (2011) is employed to increase the accuracy of Markov Chain Monte Carlo algorithm. The brief computation of antithetic variates is explained in the following section

Antithetic Variates

The technique of antithetic variates tries to reduce variance by presenting negative dependence among the pairs of repetition node. In this method, two identical samples namely X_1 and X_2 and the unbiased estimator $\hat{\theta}$ are defined as

$$\hat{\theta} = X_1 + X_2 / 2 \tag{1}$$

The variance of the samples is defined as

$$Var(\hat{\theta}) = \frac{\operatorname{var}(X_1) + \operatorname{var}(X_2) + 2Cov(X_1, X_2)}{4} \tag{2}$$

Therefore,

$$Var(\hat{\theta}) = \left[\frac{\operatorname{var}(X_1)}{2} + \frac{Cov(X_1, X_2)}{2} \right] < \operatorname{var}(X_1) / 2 \tag{3}$$

Equation (3) shows, the variance $Var(\hat{\theta})$ is smaller than $\operatorname{var}(X_1) / 2$ when, X_1 and X_2 are negatively correlated. In this case, where X_1 and X_2 are independently and identically distributed, if and only if

$$\operatorname{cov}(X_1, X_2) < 0 \tag{4}$$

Consider

$$X_i = h(U_i), \hat{X}_i = h(1 - U_i) \tag{4.7}$$

where

$$U_i = \left(U_1^{(i)}, \ldots, U_m^{(i)} \right) \tag{5}$$

and

$$1 - U_i = \left(1 - U_1^{(i)}, \ldots, 1 - U_m^{(i)}\right) For \ i = 1, 2, \ldots, n \tag{6}$$

(ie)

$$E\left(X_i\right) = E\left(\hat{X}_i\right) = \theta \tag{7}$$

Here, $h\left(U_i\right)$ is a uniform distrubution $(0,1)$ of monotone function which is described by antithetic variables approach. Further, the monotone function h, can be easily shown that the covariance of X_i & X_{i-1} is negative. Conclusively, the new estimator $Z_i = X_i + \hat{X}_i / 2$ is based on the equation (7) and is comparatively better than the usual estimator.

Locating the Sensor in the Geographical Area

In the context of many applications, it is essential to orient the nodes accurately with admiration to global co-ordinate system which relies on localization. The careful placement of sensors or uniform arrangement of sensors is a challenging task in localization process. Generally, In WSN, there are two nodes namely beacon node with its known locations and sensor node whose location has to be estimated. For example, Figure 1 shows the structure of beacon nodes and sensor nodes and the Figure contains four beacon nodes and four sensor nodes.

Figure 1. Illustration of beacon nodes and sensor nodes

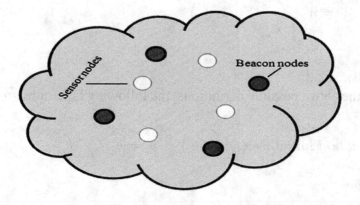

The sensor node locations are computed using Markov chain and it is defined as follows

Definition 1: Let $A = \left\{ a_{ij} \right\}_{i,j=1}^{n}$ denote an M×N matrix which allocates the sensor position in a particular area.

Here, $a_{ij} = p(X_t = s_j \mid X_{t-1} = s_i)$ represents the probability of transition from state s_i to state s_j. Where X_t is the random variable representing the sensor state at time t. The state herein refers to the region that describes the behavior of target. The moving direction of target is described by Markov chain T_j with length j

$$T_j : K_0 \rightarrow K_1 \rightarrow \dots \rightarrow K_j \tag{8}$$

Definition 2: The statistical nature of constructing the Markov chain as follows

$$p\left(k_0 = \alpha \right) = p_\alpha, p(k_j = \beta \mid k_{j-1} = \alpha) = p_{\alpha,\beta} \tag{9}$$

Where $J = 1, 2, \dots, i$ and $K_j \in \left\{ 1, 2, \dots, n \right\}$ are natural random numbers. P_α and $P_{\alpha,\beta}$ show the probability of initial chain from the node and the transition probability from state α to β. The random variable for unknown node deployment is defined by W_j and the recursion is given as

$$W_0 = \frac{h_{k_0}}{P_{k_0}}, W_j = W_{j-1} \frac{a_{k_{j-1}k_j}}{P_{k_{j-1}k_j}} \tag{10}$$

where j=1,........,i
From the above possible definitions, the following Equations are derived

$$p\left(probability\ of\ Initial\ density \right) = \{p_\alpha\}_{\alpha=1}^{n}, P_\alpha = \left| h_\alpha \right| / \sum_{\alpha=1}^{n} \left| h_\alpha \right| \tag{11}$$

$$P\left(Probability\ of\ Transition\ density\right) = \{p_{\alpha\beta}\}_{\alpha,\beta=1}^{n}\ p_{\alpha\beta} = \left|a_{\alpha\beta}\right| / \sum_{\beta=1}^{m}\left|a_{\alpha\beta}\right| \qquad (12)$$

The formulation of the initial density along with the transition density matrix supports in attaining the Optimal Monte Carlo algorithm. Consequently, the next section discusses about the localization algorithm, flow structure, environmental setup and assumptions of the algorithm.

Algorithm Description, Network Environment, and Assumptions for Localization

The algorithm for the proposed localization algorithm is given as follows based on MCL scheme:

Algorithm for Localization

Step 1: Initialize

Step 1a: For Obtaining location estimation of each nodes first initialize the localizing and non-localizing nodes that contains beacon and non-beacon nodes.

Step 1b: Initialize a global map which contains all the coordinate node information and updates the mapping structure.

Step 1c: Initialize M nodes to estimate the posterior distribution of hidden state X_k based on the observation process $Y_0,...,Y_k$ for each sensor in the network.

Step 1d: Initialize the antithetic sampling and discrete power control strategy for each sensor, based on the number of observations from M nodes.

Step 2: Update Particles using Antithetic path of Localized sensors

Step 2a: Through individual particles, the network position X_m and antithetic path are updated for every samples using antithetic MCMC (Markov Chain Monte Carlo) and for the reduction of variance of sample path process.

Step 2b: Each node initializes the no. of standard sampling to produce N paths using Markov Chain to represent a possible state.

Step 3: Enhance to Localizing Sensors

Step 3a: For individual sensor i, the antithetic variance A_i is calculated of its position estimates $\{x_{1i},...,x_{mi}\}$ as treated by individual node.

Step 3b: The lowest values of A_i are added with k sensors.

Step 4: The steps 2 to 3 are repeated, for the time being all the sensors are infused into Loc Nodes

Step 4a: The successive M network position samples are employed to illustrate a PDF through which the positions of the sensors are described.

The above algorithm explains the sequence steps for positioning the sensors. The various assumptions about the network environment of the proposed work are

1.   The nodes in the network are quasi-stationary.
2.   Nodes are left unattended at the final stage of deployment.
3.   All the nodes begin with the same amount of energy. Then, discrete power has been assigned based on RSSI and the amount of energy that optimizes based on distance.
4.   All the nodes adjust themselves, by transmitting power using radio chip CC2420.

These are some of the properties and assumptions of the localization algorithm wherein, a set of sensor localization is dispersed to a random monitoring region.

Discrete Antithetic Power Mapping

Power management technique is established by force fully charging the power utilization profile of the system, through localized computing. The profile alteration is based on the radio model by putting its sensor into power/energy states which are sufficient to meet the functionality requirements. Changing the idle sensor node into slow-down or shut down state is termed as profile change. Then, bringing such sensors into active state requires additional energy to manage the profile change. The discrete power mapping can be defined as power management strategy that minimizes the average power consumption and dissipation. To measure the effectiveness of the discrete power mapping strategy, the probabilistic scenarios like worst case and best case are considered. Also, it includes worst case delay, worst/ best power consumption and buffer size.

The discrete time steps for power mapping are explained below

Clock States

```
C:[0....1] init 0;
                [tick 1] c=00->c'  = 11
                [tick 2] c=01 ->c' = 10
                [tick 3] c=10-> c' = 01
                [tick 4] c=11 ->c' = 00
```

As shown above, the power management synchronizes with the clock on tick 1 to tick 4 and based on the choice of power state of the system. For example, in the power mapping, if the state of the system satisfies condition 1, then the probability decides that with probability p00, the SP moves to sleep, if condition 2 with the probability p01, the SP moves to idle, with p10 of condition 3, the SP moves to standby, and p11 with condition 4, the SP moves to data transfer.

Most of the nodes in WSN operate on batteries. It is important to minimize the power consumption of the entire network. The power required by each host can be classified into two categories like communication-related power and non-communication related power. The communication related power is again classified into processing power and transceiver power which are described by Minar & Mohamed (2013). A diagrammatic representation of power classification is shown in Figure 2.

Processing power is defined as the power needed to execute network algorithms and run applications and the transceiver power denotes the power used by means of the radio transceiver to communicate with one another. In clustering protocols, the sensor nodes adjust the power based on the distance (d). The energy consumption is always dissimilar for two distinct capacities

Figure 2. Power requirement classifications

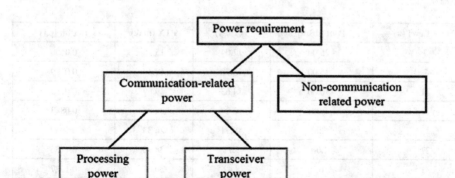

(d), when it is calculated based on distance. A unique power level is expanded for diversified values of distance, when the communicate power level of the sensor node is adjusted to discrete values, in discrete power strategy. Here, two dissimilar distances would consume similar energy consumption.

In this proposed work, the discrete power model of CC2420 is used for efficient power management. Table 1 shows the selected power level of packet size- 7 bytes (One packet consists of 100 bits) and channel rate 256Kbps.

The sensor nodes in WSN are used to adjust the power levels, according to the distance from cluster head as stated by Xuxunliu (2012).

The power level selection algorithm is shown below

Algorithm for Power Level Selection

Step 1: Receive Network Structural information from Supervisor node
Step 2: Compute distance and path loss based on RSS value(Predefined)
Step 3: Assign discrete power Level from CC2420 as shown in Table 4.1

$$\text{Discrete}_{\text{Trxs Power}} = \text{TotalP}_{\text{level}}$$

Step 4: Apply Antithetic sampling and find the sample path of

$$\{\xi_{1,\ldots,}\ \xi_M) \text{ And } \{-\xi_{1,\ldots,}\ -\xi_M)$$

Step 5: Based On the antithetic path, measure the sample size using Particle filter.

Table 1. Power output, current and transmitted energy for various power levels of CC2420

Power Level (k)	Pout [dBm]	Ix (mA)	PTX (mW)	Etx/bit [μJ]
3	- 25.00	17.04	15.15	0.0606
7	- 15.00	15.78	17.47	0.0699
11	- 10.00	14.63	19.62	0.0785
19	-5.00	12.27	22.08	0.0883
23	-3.00	10.91	26.33	0.1050
27	-1.00	9.71	28.40	0.1136
31	0	8.42	30.67	0.1227

Step 6: Check whether the measured sample size is independently distributed or not

Step 7: If yes, Covariance (Measure how much two antithetic samples change together) is negative iecov (Sample 1, Sample 2) =0 otherwise return to step 4

Step 8: Check P_{rx} <assigned discreteP_{Tx} then

Step 9: Transmit at TotalP$_{level}^{StepN}$

Step 10: Otherwise reduce the sample step size (stepN)

Step 11: Stop

Based on the algorithm, the sensor after estimating the distance, it computes the path loss and path fading. Further, the transmit power level is set as the lowest value. With the help of antithetic sampling method, every sample is obtained based on antithetic path to reduce the number of unwanted sampling for independently distributed variables. Depending on the path loss and path fading, the expected received power at the CH is calculated.

Discrete Radio Model Strategy

This section of the proposed work distillates on a radio model. It co-ordinates which power level ambience is necessary to communicate among two nodes. By means of the power level setting, the amount of diffusions is resolved based on the chip (CC2420) conditions to ensure correct estimation. The power level setting is regulated by the indication of the received signal strength which is required to transmit precisely among two nodes. In the proposed research, by using Radio Irregularity Model (RIM), path loss is calculated

$$RSSI = \text{Sending Power} - \text{Path loss} + \text{Fading} \qquad (13)$$

The process of controlling the transmission power is discussed in Figure 3. In the initial step, the node connection is checked as false or true. It means that two motes cannot communicate. Further, the selected sensor node level which T uses to transmit the packet to receiver. The for-loop in step 2 initiates with the lowest available power level and works up to the highest. In step 3, the transmission power level for the transmitter mote is attuned formerly by estimating the RSSI and accomplishes antithetic sampling with the set power level in step 4. As a result, the sample size and communication cost are reduced.

Process of Controlling the Discrete Transmission Power Based on Antithetic Markov Process

See Figure 3.

RESULTS AND DISCUSSION

Here, the proposed localization scheme (AMCMC) is measured through simulations. The algorithm is done in Matlab. The effectiveness and the efficiency of the proposed work are measured through parameters like Localization accuracy, Localization error, computational time and sampling efficiency.

Localization Accuracy

It is the key parameter in most of the applications of WSN. It depicts the localization accuracy percentage at different node densities. The node density is defined as the number of nodes in unit area. The Table 2 shows the accuracy percentage of Monte Carlo Localization (MCL), Adaptive Monte Carlo Localization (AMCL) and Antithetic Markov Chain Monte Carlo Localization (AMCMC) at different node densities ranging from 0-45. Here, the average localization accuracy is measured based on the number of unknown nodes which acquires its location over the total number of sensor nodes.

Figure 3. Power transmission control procedure

```
Step 1: Node Connected  ←   True (or) False
Step 2: for each Node_j ∈ P_Max do
Step3 : Power level < - j
Step 4: Assign P_Level = Node_j & Perform Antithetic sampling
Step 5: if (Predefined RSSI and sample size _ min) then
Step 6: connected <- True
Step 7: Fix the sample size & Measure the Position of the sensor
Step 8: break
Step 9: end if
Step 10: end for
Step 11: return Sample size & Communication cost (T.Power level, Sample Size)
```

Table 2. Localization accuracy of MCL, AMCL and AMCMC at different node densities

Node Density (Nodes/ Meter Square)	Accuracy%		
	MCL	AMCL	AMCMC
0-5	60	65	70
6-10	57	64	73
11-15	55	63	75
16-20	52	61	78
20-25	50	60	78
26-30	49	59	83
30-35	49	55	84
35-40	45	52	85
41-45	45	50	94

The Figure 4 shows the statistical representation of the average accuracy obtained at different node densities by the existing method and the proposed method. The accuracy percentage of the Monte Carlo Localization (MCL) gradually decreases, when the node density increases. At the start up node density, 60% accuracy is obtained, when the density of the node increases, the accuracy is decreased to 45% at the node level 41-45.

Also, in Adaptive Monte Carlo Localization, 65% accuracy is achieved at the initial nodes and it gets drop down to 50%, when the node density increases to 41-45. The lack of Discrete Power Control (DPC) strategy is the main reason for the reduction in accuracy level, when the network size increases. In the proposed Antithetic Markov chain Monte Carlo (AMCMC), the DPC strategy and Antithetic variates are used to obtain the high localization accuracy.

The AMCMC algorithm attains 70% of accuracy at the preliminary network size and the accuracy is slightly increased at the successive density. At the final point, the accuracy is increased to 94% wherein nearly 24% of accuracy is increased.

The accuracy obtained by AMCMC is 10% higher than the MCL and 5% higher than the AMCL at the node density 0-5. At the node level 41-45, the accuracy is 49% higher than MCL and 44% higher than AMCL.

Here, in AMCMC the accuracy is maximized, although the network size increases, whereas in the existing algorithm, the accuracy is minimized, when the node density increases.

Figure 4. Localization accuracy for MCL, AMCL and proposed AMCMC algorithm based on node density

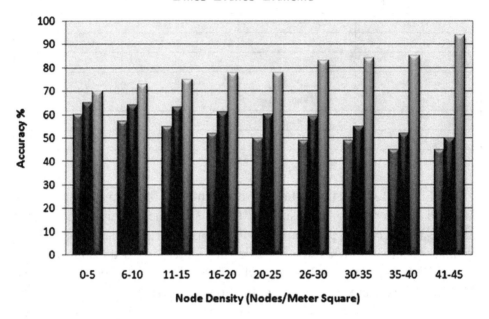

*For a more accurate representation see the electronic version.

Localization Error

Average Error range is one of the performance factors for measuring the localization algorithm performance of the research phase. The localization error in this work refers to the probability of unidentified node's position in localization process. Normally, the error ranges between 0-1 (Error Probability) in terms of probability measurement. When the error is minimized, then the performance of the algorithm is convincing otherwise, some modifications are needed to enhance the performance. The Table 3 gives a brief description about the Error range factor of the proposed AMCMC and the existing algorithms, MCL and AMCL.

The average error of MCL technique ranges from 0.4-0.55 and in AMCL technique, the error ranges between 0.35-0.5 from the initial level to the highest node level. The error ranges between 0.3-0.06 in the proposed AMCMC.

The error range is minimized here. Whereas, in the existing algorithm, as the node density increases the error range is increased.

Table 3 demonstrates the comparison between the existing and the proposed localization techniques in terms of error range. The x-axis of the figure shows the various node densities and the y-axis of the figure shows the error range. The error range is minimized considerably in the proposed system compared to the existing system.

The performance of the proposed technique is quite satisfying compared to the existing MCL and AMCL techniques. The AMCMC technique uses one- hop distance estimation which is less sensitive to the obstacles. It uses only local information for effective localization process.

The Figure 5 shows the simulation results of mobile node speed of 2 m/s. It is equivalent to 0.25 *r/unit time*. The results obtained are compared to the results related to the node speed of 8 m/s which is equivalent to 1 *r/unit time* for the network sizes of 20, 30 and 40. The irregular density does not affect the performance of the proposed AMCMC algorithm very much. It uses one-hop distance estimation that helps to update and produce the outcome that is less sensitive to obstacles. It works by utilizing only local information.

Computational Time and Sampling Efficiency

Time complexity is the most important performance factor for analyzing the Localization progression of this research phase. The Table 4 gives a brief

Table 3. Different node density and error range for AMC, AMCL and AMCMC algorithm

Node Density (Nodes/Meter Square)	Error Range (Probability)		
	MCL	AMCL	AMCMC
0-5	0.4	0.35	0.3
6-10	0.43	0.36	0.27
11-15	0.45	0.37	0.25
16-20	0.48	0.39	0.22
20-25	0.5	0.4	0.22
26-30	0.51	0.41	0.17
30-35	0.51	0.45	0.16
35-40	0.55	0.48	0.15
41-45	0.55	0.5	0.06

Figure 5. Error computation for MCL, AMCL and proposed AMCMC algorithm based on node density

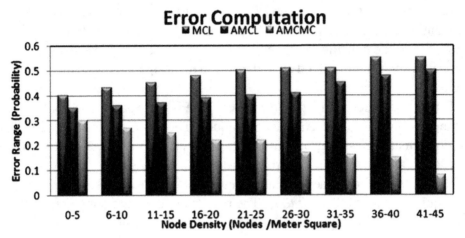

For a more accurate representation see the electronic version.

Figure 6. Localization error at different speed

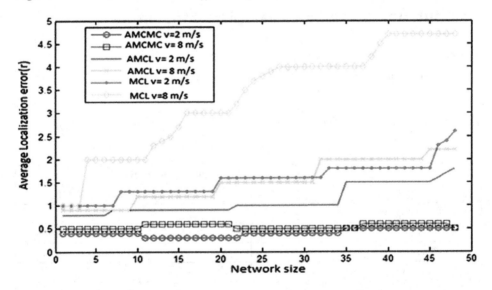

description about the Computational time performance factor between the proposed AMCMC technique and the existing MCL and AMCL techniques.

Table 4 depicts the time taken by the methodologies related with the process of localization at different node densities and it is expressed in seconds. The

Table 4. Different node density & computational time for AMC, AMCL with proposed AMCMC algorithm

Node Density (Nodes/Meter Square)	Computational Time (s)		
	MCL	AMCL	AMCMC
0-5	5	7	2
6-10	5	17	3
11-15	8	28	5
16-20	12	32	5
20-25	18	37	6
26-30	23	40	7
30-35	28	47	7
35-40	36	62	8
41-45	42	76	8

time is derived for assessing the speed of the process at different network sizes. Consequently, when number of nodes increases in a WSN, the computational time taken to localize the nodes also increases. The computational time of AMCMC algorithm is much less than that of MCL and AMCL.

Figure 7 shows the computational time comparison of the proposed AMCMC algorithm and the existing MCL and AMCL techniques. Basically, the MCL technique utilizes 5s to 42 s and AMCL takes 7s to 76 s for localizing the nodes from 0-45 densities.

In the existing localization technique, the computational time is increased from time–to–time, as the node density increases. The computational time is literally increased or same, as the density increases in AMCMC technique.

Both MCL and AMCL cannot provide effective results, as they take more time for localizing the nodes and also sometimes, the results are outdated, due to time complexity. On the other hand, the AMCMC technique ensures less execution time with an increase of the network size. Thus, the AMCMC provides accurate results. The proposed method provides accurate results at shorter time intervals compared to the existing technique.

The Table 5 shows the candidate sampling of the proposed AMCMC algorithm which is compared against the candidate sampling of existing methods MCL, AMCL and IMCL. In general, sampling efficiency is improved based on number of candidate sampling, which means less candidate sampling and consequently, less computational cost. when, as the number of beacon nodes increases in a WSN, the candidate sampling for localization process

Figure 7. Computational time for MCL, AMCL and proposed AMCMC algorithm based on node density

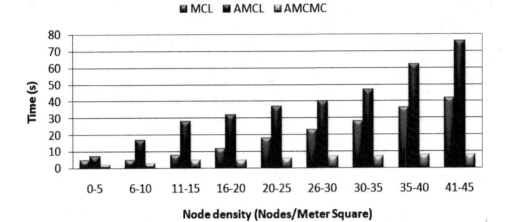

*For a more accurate representation see the electronic version.

Table 5. Analysis of sampling efficiency based on number of beacon nodes with existing Monte Carlo methods

Number of beacon nodes	2	4	6	8
Candidate sampling for MCL	580	565	550	510
Candidate sampling for AMCL	510	480	425	320
Candidate sampling For IMCL(Iterative MCL)	450	425	400	360
Candidate sampling for proposed AMCMC	390	320	310	290

also increases. In AMCMC, the increase is much lesser than that of MCL, AMCL and IMCL. Hence, the proposed method achieves higher sampling efficiency and Lower computational cost.

As a result, the proposed AMCMC is efficient in terms of accuracy, error, computational time and sampling efficiency, as the number of node increases. AMCMC is better than the existing technique and it provides sufficient accurate results and low computational error within the shorter time span. Thus, the AMCMC is deliberately designed to ensure the objectives.

CONCLUSION

The power plays a major role in WSN and it is generally supplied by the battery. It is essential for computation purpose. The main goal is to make the devices to utilize limited power. It plays a vital role in deployment of nodes. Choosing the devices is also important for both the lifespan of the nodes and cost of the power. In the proposed work, the power concept is focused for greater efficiency in terms of energy consumption, lifespan, cost, accuracy and good positioning. This paper is mainly focused on improving the process of localization using antithetic variates for variance reduction and discrete power control strategy. The research findings promise better performance of the proposed AMCMC than the existing MCL and AMCL. The proposed AMCMC, algorithm increases the accuracy by computing the dominant eigen pair of a matrix. The Monte-Carlo-based approximation analyses the sampling rate of each transmission and energy based distance estimation. From the simulation results, it is proved that the proposed algorithm increases the accuracy of localization and minimizes the estimation errors. It has also reduced the energy consumption and over head. Although the enhancement moves the process to the next level, the system has to be improved by considering the inadequacy in obtaining optimal results.

REFERENCES

Abu Znaid, I., Idris, M. Y. I., Wahab, A. W. A., Qabajeh, L. K., & Mahdi, O. A. (2017). Low communication cost (LCC) scheme for localizing mobile wireless sensor networks. *Wireless Networks*, *23*(3), 737–747. doi:10.100711276-015-1187-6

Alhashimi, Hostettler, & Gustafsson. (2014). An improvement in the observation model for montecarlo localization. *Proceedings of the eleventh international conference on informatics in control, Automation and robotics*, 498-505.

Barberis, A., Barboni, L., & Valle, M. (2007). Evaluating energy consumption in wireless sensor networks applications. *Proceedings of the Tenth Euro Micro Conference on Digital System Design Architectures, Methods*, 455-462. doi: 10.1109/DSD.2007.4341509

Bennani, K., & El Ghanami. (2012). Particle swarm optimization based clustering in wireless sensor networks: The effectiveness of distance altering. *Proceedings of the International Conference on Complex Systems*, 1-4. doi: 10.1109/ICoCS.2012.6458564

Cai & Wonham. (2010). Supervisor localization: a top-down approach to distributed control of discrete-event systems. *IEEE Transactions in Automatic Control, 55*(3), 605-618.

Dasgupta, K., Kalpakis, K., & Namjoshi, P. (2003). An efficient clustering based heuristic for data gathering and aggregation in sensor networks. *Proceedings of the International Conference on Wireless Communications and Networking*, 1948-1953. doi:10.1109/WCNC.2003.1200685

El Aaasser, M., & Ashoar, M. (2013). Energy aware classification for wireless sensor networks routing. *Proceedings of the fifteenth international conference on advanced communication technology*, 66-71.

Halder, G., & Ghosal, A. (2016). A survey on mobility assisted localization techniques in wireless sensor networks. *Journal of Network and Computer Applications, 60*, 82–94. doi:10.1016/j.jnca.2015.11.019

He, T., Huang, C., Blum, B. M., Stankovic, J. A., & Abdel Zaher, T. (2003). Range-free localization schemes for large scale sensor networks. *Proceedings of the Ninth Annual International Conference on Mobile Computing and Networking*, 81-95. doi:10.1145/938985.938995

Jeong, Culler, & Hyukoh. (2007). Empirical analysis of transmission power control algorithms for Wireless sensor networks. *Proceedings of the fourth international conference on Networked Sensing Systems*, 27-34. doi:10.1109/ INSS.2007.4297383

Kroese, D. P., Taimre, T., & Botev, Z. I. (Eds.). (2011). *Variance reduction, in hand book of montecarlo methods*. John Wiley & Sons.

Kurecka, A., Konechy, J., Prauzek, M., & Koziorek, J. (2014). Monte Caro based wireless node localization. *Elektronika Elektro Technika, 20*(6), 12-16.

Naraghi Pour, M., & Rojas, G. C. (2013). Sensor network localization via distributed randomized gradient descent. *Proceedings of the IEEE Military Communications Conference*, 1714–1719. doi:10.1109/MILCOM.2013.290

Neufeld, Schuurmants, & Bowling. (2015). Variance reduction via Antithetic Markov chains. *Proceedings of the eighteenth international conference on Artificial intelligence*, 708-716.

Niu, Huan, & Chen. (2016). NMCT: a novel Monte Carlo-based tracking algorithm using potential proximity information. *International Journal of Distributed Sensor Networks*, 481-495.

Paul & Sato. (2017). Localization in Wireless Sensor Networks: A survey on Algorithms, Measurement Techniques, Applications and Challenges. *Journal of Sensor and Actuator Networks, 6*, 1–23.

Praveena, Vennila, Kumar, & Pravin. (2014). An enhanced localization fusion scheme for mobile sensor networks using Monte Carlo localization. *Australian Journal of Basic and Applied Sciences, 18*(7), 426-434.

Rao, W., Zhu, H., & Lu, Z. (2009). An advanced distributed MDS-MAP algorithm for WSNs. *Proceedings of the International Conference on E-Business and Information System Security*, 1-5. 1 doi:0.1109/EBISS.2009.5137917

Sahoo, R. R. (2014). *Design and application of iterative monte carlo localization for mobile wireless sensor networks based on MCL.* Computer and Radio Engineering.

Shang, Y., Shi, H., & Ahmed, A. A. (2004). Performance study of localization methods for ad hoc sensor networks. *Proceedings of the IEEE International Conference on Mobile Ad-hoc and Sensor Systems*, 184–193. doi:10.1109/MAHSS.2004.1392106

Singh, Tripathi, & Singh. (2011). Localization in Wireless Sensor Networks. *International Journal of Computer Science and Information Technologies, 2*(6), 2569-2572.

Singh, Bhoi, & Khilar. (2017). Geometric Constraint Based Range Free Localization Scheme for Wireless Sensor Networks. *IEEE Sensors Journal, 17*, 5350-5366.

Teng, G., Zheng, K., & Dong, W. (2011). MA-MCL: Mobile-Assisted Monte Carlo Localization for wireless sensor networks. *International Journal of Distributed Sensor Networks.*

Wang & Zhu. (2009). Sequential monte carlo localization in mobile sensor networks. *Journal of Wireless Networks, 15*(4).

Wang & Zhu. (2009). Sequential monte carlo localization in mobile sensor network. *Wireless Networks, 15*(4), 481–495.

Wu, X., Wu, H., Liu, Y., Zhang, G., & Xing, J. (2015). Localization algorithm for mobile nodes in wireless sensor networks based on H_∞ filtering and dynamic node model. *Journal of Communication, 10*(6), 415–422.

Xu. (2012). A survey on clustering routing protocols in wireless sensor networks. *Sensors, 12*(8), 11113-11153.

Zhang, S., Cao, J., Chen, L.-J., & Chen, D. (2010). Accurate and energy-efficient range-free localization for mobile sensor networks. *IEEE Transactions on Mobile Computing, 9*(6), 897–910. doi:10.1109/TMC.2010.39

Zheng, Liu, Yang, Chen, Martin, & Li. (2014). A study of localization accuracy using multiple frequencies and powers. *IEEE Transactions on Parallel and Distributed Systems, 25*(8), 1955-1965.

Zhou, G., He, T., Krishnamurthy, S., & Stankovic, J. A. (2004). Impact of radio irregularity on wireless sensor networks. *Proceedings of the Second International Conference on Mobile systems, Applications and Services*, 125–138. doi:10.1145/990064.990081

Chapter 5
LEACH-VD:
A Hybrid and Energy-Saving Approach for Wireless Cooperative Sensor Networks

Proshikshya Mukherjee
KIIT University (Deemed), India

ABSTRACT

Wireless sensor networks act as an important role in the wireless communication area because of its properties, its intelligence, cheaper costs, and its smaller size. Multiple nodes are required for coperative communication, the low energy adaptive clustering hierarchy and LEACH-Vector Quantization are used for cluster and active cluster head formation. Further, Dijkstra Algorithm is used to find the shortest path between the active CHs and high-energy utilization, respectively. The main issue of inter-cluster communication is carried out in earlier work using LEACH and LEACH-V protocols. The chapter illustrates the LEACH-Vector Quantization Dijkstra protocol for shortest path active CH communication in a cooperative communication network. In the application point of view, LEACH-VD performs the lowest energy path. LEACH-V provides the intra-cluster communication between the cluster head, and using Dijkstra Algorithm, the minimum distance is calculated connecting the active cluster heads, which creates the shortest path results using an energy-efficient technique.

DOI: 10.4018/978-1-5225-9004-0.ch005

INTRODUCTION

Wireless sensor network (WSN) it sometimes called wireless actuator network. The WSN is specially consists of distributed autonomous sensors which monitored environmental or physical conditions like temperature, pressure, sounds, etc. and passes their sensed information through the path to the main location (Yadav and Sunitha, 2014). These paths are made by using routing. The routing is a process to create a path between a source node to the destination node. Various types of routing protocol are used for the communication purpose. These routing protocols are affected in WSN by several exigent factors like throughput, scalability, bandwidth utilization, network lifetime, etc. Routing protocol can be classified five ways like, according to the ways, according to the way of establishing the routing paths, according to the network structure, according to the protocol operation, according to the initiator of communications and according to how a protocol selects a next hop on the route of forwarded message (Li et. al, 2011).

When the energy efficiency and stability is needed, then the cluster based routing protocols are used. In hierarchical cluster based is more energy efficient because high energy nodes are randomly data selected for processing and sending information and low energy nodes are used for sensing and sending data to the clustered (CH). In this process network life time and stability of the network are increased. LEACH (Low Energy Adaptive Clustering hierarchy) is one of the cluster based hierarchical protocol. This protocol is self-adaptive and self-organizing. LEACH (Heinzelman, Chandrakasan, and Balakrishnan, 2002) with vector quantization is used for the intra-cluster communication between the cluster head (CH). This reason less amount of power is used than the LEACH. So, that the network lifetime is increased than the LEACH. Dijkstra's algorithm is used for the shortest path between the nodes in a graph. The proposed methodology is the combination of LEACH-V with Dijkstra's algorithm which show that the optimum path between the active cluster head (CH) and amount of energy utilized in this path. This paper flow is: LEACH protocol overview and its energy utilization, then LECH-V protocol and its inter-cluster communication and its energy utilization. The next section is the Dijkstra's Algorithm and where the optimum shortest path between the active cluster-head (CH) network. Then the compression among them. Then finally, the conclusion and future work of the proposed methodology in WSN.

OVERVIEW OF LEACH

LEACH is an energy conserving cluster based routing protocol. The whole networks are divided into several clusters. In several round clusters are broken in the run time. In each and every round cluster head (CH) is made by their predefine criteria. Every node has equal probability to make a cluster head (CH) (Fu et. al, 2013).

Figure 1 shows that the several numbers of nodes. The nodes have created a group of cluster. Every cluster has a head node called cluster head. All cluster heads are directly connected to the base station. In LEACH protocol each round has two stages i.e. steady state and setup state. Steady stages always larger than the setup stage. Figure 2 shows the LEACH TDMA procedure.

'n' stands for the node which is a random variable between 0 and 1. is the predefined threshold value. When node 'n' is lesser than the threshold value then the cluster head is created. Otherwise the nodes are called common nodes. Equation (1) shows that the threshold value of LEACH protocol,

Figure 1.

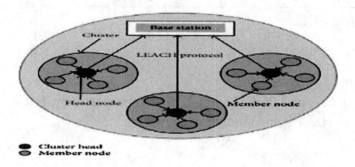

Figure 2.

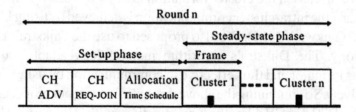

$$T(n) = \frac{p}{1 - p\left(r \bmod \frac{1}{p}\right)} \quad \forall n \in G$$

$$T(n) = 0 \quad \forall n \notin G \tag{1}$$

OVERVIEW OF LEACH-V

In this method multiple cluster heads (CH) communicated using vector quantization method fond the minimum Euclidean distance (storage path). This process consumes the power. Initially WSN fixed the node. Fixed a threshold value and created a cluster head (CH) by using the LEACH algorithm. After creating clusters and cluster head (CH) vector quantization is used for minimum distance between the cluster head.

The major objective is calculating the spectral distortion of LEACH-V (Samanta et. al, 2016). For the ith position spectral distortion is form equation (2)

$$SD_i = \sqrt{\frac{1}{(f_2 - f_1)} \sum_{f_1}^{f_2} \frac{p(f_2) - p(f_1)}{1 - \left[(p(f_2) - p(f_1)) * \left(r \bmod \frac{1}{(p(f_2) - p(f_1))}\right)\right]}} \tag{2}$$

where, f_1 and f_2 are 3.36 GHz and 3.56 GHz respectively. f_1 and f_2 are the frequency in Hz. 'r' is the number of round.

PROPOSED METHODOLOGY

As we have discussed the cluster formation is achieved through the LEACH protocols and the intra-cluster communication is achieved by using LEACH-V (Barai and Gaikwad, 2014). We also proposed to use the Dijkstra's algorithm methodology. The Dijkstra's algorithm methodology adapts the shortest distance as its major and thus effective in communication by using the energy comparatively. So combining the above three mechanism our proposed method "LEACH-VD"(LEACH -Vector Quantization with Dijkstra's algorithm) to be implemented in a co-operative network. Initially for WSN, an area dimension has been set with a fixed number of nodes. We have taken that the various participating nodes are randomly distributed working with two different unlicensed frequencies. Here we implemented LEACH by setting a threshold for CH for formation of cluster vector.

After formation of clusters and their respective CH, VQ is implemented for analysing the minimum distance between the CHs. After formation of CH, Dijkstra's algorithm is used for the shortest distance between the active CHs.

Modified Algorithm

Step 1: LEACH (To find active nodes for preparing cluster energy efficiently)
Step 2: LEACH-V (To find active CHs and the shortest distance between all of them from the output of Step 1)
Step 3: LEACH-VD (To find the shortest path from a particular CH to another CH based on the all the active paths obtained from Step 2)

Results And Discussion

In LEACH algorithm, we are considering 30 nodes. In the distance is 6X6 km. Node id is declared the each node position. Figure 3 shows that the Node ID vs Energy levels graph.

Figure 3. Energy usage using LEACH (30 nodes); average energy utilization per node (in percentage): 33.5%

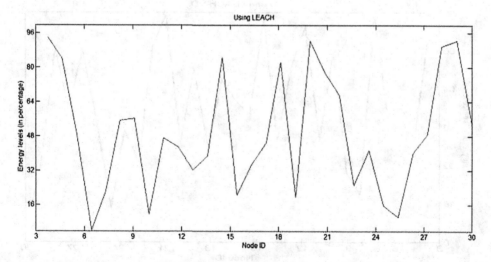

Figure 4. Energy usage using LEACH-V (30 nodes); average energy utilization per node (in percentage): 41.5%

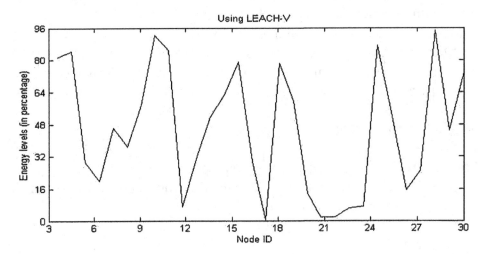

In LEACH- V algorithm, we are considering 30 nodes. In the distance is 6X6 km. Node id is declared the each node position. Figure 5 shows that the Node ID vs Energy levels graph. In this algorithm we are adding the vector quantization.

Figure 5. Energy usage using LEACH-VD (30 nodes); average energy utilization per node (in percentage): 58.3%

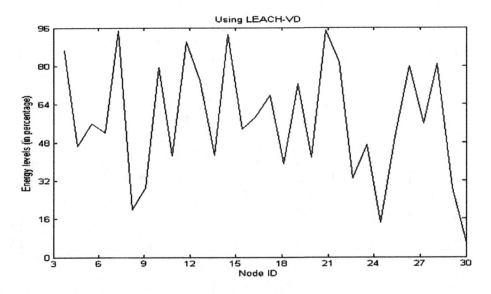

Figure 6 shows that the Node ID vs Energy levels graph. In this algorithm we are adding the vector quantization.

Figure 7 shows that using LEACH algorithm average energy utilization is 33.5%. Same as Using LEACH-V average energy utilization is 41.5%. But using Dijkstra's algorithm average energy utilization is 58.3%. So the energy utilization is increased in LEACH-VD. Figure 4 shows that which path optimum path in dijkstra's algorithm.

In LEACH algorithm, we are considering 30 nodes. In the distance is 6X6 km. Node id is declared the each node position. Figure 7 shows that the Width vs length graph. In this graph black line shows that the cluster head node connected each other . The red line shows that the optimum path of the graph using Dijkstra's algorithm.

CONCLUSION

The proposed technique, extends the LEACH protocol to LEACH- VQ-Dijkstra's protocol (LEACH-VD) for energy utilization in the lowest energy path. Here, LEACH provides the optimum cluster size, whereas by using VQ the minimum distance is calculated between multiple cluster head. In

Figure 6. Comparison of graph

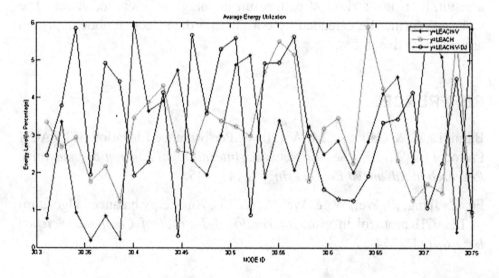

Figure 7. LEACH-VD (Assuming starting node 1 and en node 30)

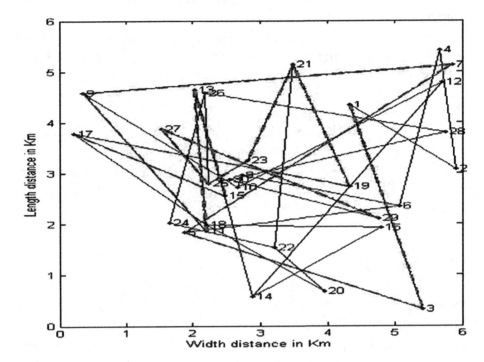

*For a more accurate representation see the electronic version.

Dijkstra's shows the minimum distance between the active cluster heads. As a result, it creates a shortest path result in energy efficient technique. The work can further be extended by considering the selective node approach among the active nodes.

REFERENCES

Barai, L. Y., & Gaikwad, M. A. (2014). Performance Elevation of LEACH Protocol for Wireless Sensor Network. *International Journal of Innovative Research in Advanced Engineering, 1*, 141–145.

Fu, C., Jiang, Z., Wei, W., & Wei, A. (2013). An energy balanced algorithm of LEACH protocol in wsn. *International Journal of Computer Science Issues, 10*(1), 354-59.

Heinzelman, W., Chandrakasan, A., & Balakrishnan, H. (2002). An application-specific protocol architecture for wireless microsensor networks. *IEEE Transactions on Wireless Communications*, *1*(4), 660–670. doi:10.1109/TWC.2002.804190

Li, Y., Yu, N., Zhang, W., Zhao, W., You, X., & Daneshmand, M. (2011). Enhancing the performance of LEACH protocol in wireless sensor networks. *IEEE INFOCOM Workshop on M2MCN*, 223-28.

Samanta, T., Mukherjee, P., Mukherjee, A., Swain, T., & Dutta, A. (2016). LEACH-V: A solution for intra-cluster cooperative communicationin wireless sensor network. *Indian Journal of Science and Technology*, *9*(48). doi:10.17485/ijst/2016/v9i48/100619

Yadav, L., & Sunitha, C. (2014). Low energy adaptive clustering Hierarchy in wireless sensor network (LEACH). *International Journal of Computer Science and Information Technologies, 5*(3), 4661-64.

Chapter 6

An Integrated GIS and Knowledge–Based Automated Decision Support System for Precision Agriculture Using IoT

Raja Lavanya
Thiagarajar College of Engineering, India

G. Vinoth Chakkaravarthy
Velammal College of Engineering and Technology, India

P. Alli
Velammal College of Engineering and Technology, India

ABSTRACT

Agriculture is a sector that holds great promise to Indian economic growth. Production in rural Tamil Nadu is extremely low due to unscientific farming practices. The major challenges faced in Tamil Nadu agriculture are crop mapping, yield prediction, quality of food produced, irrigation management, variable rate fertilizer and pesticide due to lack of technical knowledge. Precision agriculture (PA) rules out all drawbacks of traditional agriculture. The main objective of the chapter is to enhance the productivity of rural Tamil Nadu in order to meet the growing demands of our country's food supply chain

DOI: 10.4018/978-1-5225-9004-0.ch006

INTRODUCTION

Agriculture is the art of science for cultivating the soil, producing crops using different preparation methods and technologies and marketing the resultant products produced in the farming. India is an agriculture based country. The farmers are referred as ecosystem engineers as they use many different methods and techniques for increasing the production. It's crucial to maximize agriculture resources in a sustainable manner. This is achieved by continuously monitoring and collecting data of crops and its major factors affecting the productivity of the crop such as weather, soil and air quality, crop maturity. An experimental test bed has to be designed and various types of sensors can be placed throughout the fields which are used to measure the needed parameters (Haider, Rosdiadee, Sadik, Aqeel & Mahamod, 2017).The proposed work adopts different domains like database management systems, Data analytics, Data mining, Internet Of Things, Wireless sensor Networks and Image processing and Artificial Neural Network. The physical real time data can be captured through different sensors and with the help of wireless sensor networks, the data will be uploaded in the cloud. The uploaded data will be properly maintained through database management systems in order to have the well-organized data. Then data mining algorithms will be applied to support the decision making process. During the entire cultivation period, the proper monitoring can be done. Also a recommended system has to be developed to predict future conditions and help farmers to make proactive smart decisions. At the end of harvest, proper evaluation will be done and revision plan will be generated for further improvement.

Objectives of the Work

1. To help the farmers in Rural Tamil Nadu by providing optimized solution for improved agricultural production with the adaptation of Precision Agriculture in order to meet the growing demands of our country's food supply chain.
2. To optimize the inputs for agricultural production with respect to the nature of the agricultural land through Integrated GIS and Knowledge Based Automated Decision Support System in order to maximize the agriculture resources in a sustainable manner.

3. To improve monitoring methods which include crop mapping, yield prediction, quality of food produced with the help of efficient inter disciplinary computing technologies.

Problems Intended to Be Addressed by the Proposed Work

- Agricultural productivity in India is much lower than it should be. But dispersed rural populations and difficulty in monitoring agricultural extension agents reveal the limitations of current methods of information dissemination.
- Lack of knowledge in the determination of the effect of soil type, landscape position, nutrient level, fertility treatments, disease and weeds of crops.
- Lack of skills in water irrigation management.
- Lack of digital analysis to detect crop disease.
- Lack of decision making systems to optimize fertilizers and pesticide inputs and to reduce environmental risk of the production .
- Lack of Identification of equipment development and technology transfer needs.

One of the biggest reasons for people to drift away from agriculture is that people do not find profit from farming. (Anup, Ojas, Deepika, & Sushmita, 2016) The adaptation of technological status of traditional agriculture of India is very limited when compared with the international status. As said India has the second largest area of fertile land but lacking in adopting new advanced computing technologies in the agricultural domain. This system will guide the farmers to use the advanced technology monitoring methods which include crop mapping, yield prediction, quality of food produced, irrigation management, variable rate fertilizer and pesticide to increase the productivity.

LITERATURE SURVEY

Precision agriculture concept is spreading rapidly in developed countries as a tool to fight the challenge of agricultural sustainability. With the progress and application of information technology in agriculture and IT revolution in developed countries like USA, Australia and China etc., precision agriculture has been increasingly gaining attentions worldwide .At the same time, among the developing countries, Argentina, Brazil, India, Malaysia, and others have begun to adopt some PA components, especially on research farms, but the adoption is still very limited. Constant watch on the present status of a technology helps to identify adoption trend and converge research effort in the same direction. Therefore, exhaustive review on the status of PA in developing countries is very important.

Overview of PA Technology in the Developed Countries

Positioning System

Global Positioning System (GPS), developed by US Department of Defense, is based on constellation of 24 satellites in 6 orbital planes with orbit period of 12 h at 55° inclination with an altitude of 12,500 mile above the earth. Similar positioning system is Russian GLONASS positioning system. A European Global Navigation Satellite System (GNSS), Galileo is also under processing. Each GPS satellite orbits, has an atomic (caesium vapour) clock, which is an international time standard. Time synchronization of the coded signals transmitted by the satellites provides the basis of the system, which allows a ground level receiver to compute its range from each satellite currently in view and hence via the measured range to three or more satellites-to compute its position on the earth's surface . A ground based network of reference stations providing correction data and other information for the users is necessary for real time, high accuracy (down to the 1 cm level) GPS positioning.

Yield Mapping

Yield is decisive pointer of variation of different agronomic parameters in different parts within the field. So mapping of yield and correlation of that map with the spatial and temporal variability of different agronomic parameters helps in development of next season's crop management strategy .Present yield monitors measure the volume or mass flow rate to generate time periodic record of quantity of harvested crop for that period .Grain yields are measured using four types of yield sensors-impact or mass flow sensors, weight-based sensors, optical yield sensors and γ-ray sensors. Most major agricultural equipment companies provide optional yield-mapping systems for their combine harvesters .

Remote Sensing

Remotely sensed information, containing electromagnetic data of crop can provide information useful for agronomic management. This type of information is cost effective and can be very useful for site-specific crop management programs . Mapping of weeds against bare soil for row crops at early stages of seedlings has been carried out successfully. As a general theory resolutions should be less than the minimum expected size of weed patch . Synthetic Aperture Radar (SAR) images from satellite like ESR2 can penetrate cloud cover. So combination of SAR images with multi-spectral images can improve spatial crop management decision .

Soil and Crop Sensing

Collection and analysis of soil and plant tissue sample in laboratory is time and cost intensive. In recent years many instruments have been invented based on direct contact and proximate remote sensing technology . (Nikesh & Kawitkar, 2016). Many experiments have been already proved close relationship between crop yield and soil Electrical Conductivity (EC). Combining a soil EC probe with an automated penetrometer, soil subsurface can be mapped . A penetrometer equipped with a near-infrared reflectance sensor measured soil penetration resistance as well as and . Ion Selective Field Effect Transistor (ISFET) has been proved superior to Ion Selective Electrode (ISE) in several respects like high signal/noise ratio, fast response etc.

Variable Rate Technology (VRT)

Existing field machinery with added Electronic Control Unit (ECU) and onboard GPS can fulfill the variable rate requirement of crop. Bennett and Brown developed a direct nozzle injection system for herbicide application. An innovative distributed control system between individual sensors and actuators, a supervising controller and a navigation system was designed and installed to control spray droplet size and application rate for agricultural chemicals . Spray booms, spinning disc applicator with ECU and GPS have been used for patch spraying . Granular applicators equipped with VRT have gained popularity in recent years as a result of increased interest in variable-rate application. Swisher designed an optical sensor to measure flow rates of granular fertilizer in air streams for feedback control of a variable-rate spreader.

Proposed Framework

The proposed framework aims to bridge the computing and communication systems with the physical world whose operations are monitored, coordinated, controlled and integrated through different inter disciplinary technologies.

It adopts different domains like Database Management Systems, Data analytics, Data mining, Internet of Things, Wireless Sensor Networks and Image processing and Artificial Neural Network. The physical real time data can be captured through different sensors and with the help of wireless sensor networks, then the data will be uploaded in the cloud. The uploaded data will be properly maintained through database management systems in order to have the well-organized data. Then data mining algorithms will be applied to support the decision making process. During the entire cultivation period, the proper monitoring can be done. At the end of harvest, proper evaluation will be done and revision plan will be generated for further improvement.

The important constituent part of the IoT, sensor network, enables us to interact with the real world objects. In this proposed work, we deal with the sensor network design that enables connecting agriculture to the IoT (Ojas, Anup, Deepika, & Sushmita, 2015). The connection sets up the links among agronomists, farms and thus improves the production of agricultural products. It is a comprehensive system designed to achieve precision in agriculture.

This system will include the intelligent system which will take the decisions or actions according to the conditions prevailing. So that the farmers' interaction with the system will be minimized which will lead to less human efforts for the monitoring. The proposed work also uses image processing techniques with Artificial Neural Network to support disease identification.

The likely impacts foreseen in this subject area by the proposed work are:

1. The proposed work can provide significant economic benefits to the farmers and also provide environmental benefits.
2. The proposed work helps the farmers to identify the problems and opportunities what they were unaware of.
3. This work also helps the farmers to make logical decisions towards crop selection, crop monitoring, irrigation management and using pesticides for identified diseases.

The outcome of this work is to deploy the developed solution in the agricultural lands. Once it has been successfully deployed, ultimately farmers of our country are benefitted, which in turn helps our Government .

The enablement path of the project would be as follows:

Historical Data collection → Data maintenance system → Data analysis and mining
→ Crop monitoring through smart irrigation system
→ Image analysis collected through drones for crop disease identification
→ Alert system for climatical disasters → Evaluation and revision plan

METHODOLOGY

The proposed work tries to alleviate the different issues of traditional agriculture in both economic perspective and environmental perspectives. The proposed work adopts the following modules:

1.   Historical Data collection
2.   Data maintenance system.
3.   Data analysis and mining
4.   Crop monitoring through smart irrigation system.
5.   Image analysis collected through drones for crop disease identification.
6.   Alert system for climatical disasters
7.   Evaluation and revision plan

1. Historical Data

Historical data is the collection of geographical information about agricultural lands which is used for decision making process. These data are collected through various sources like grid soil samples, detailed soil mapping, topographic maps, soil texture maps, environmentally sensitive areas and others. All the data are georeferenced as it is collected. The proposed work uses the different kinds of sensors and wireless network using Zig bee for data collection and transfer.

The proposed system consists of the following components for data collection:

1.   Sensors to detect parameters of soil like soil moisture, temperature, humidity, nitrogen, potassium and phosphorus
2.   Arduino Uno R3 microcontroller (or) Raspberry pi
3.   Zigbee

4.   Server

Sensors are the device which converts physical parameter into the electrical signal. The system consists of 6 different sensors to collect data of the different parameter of the soil and atmosphere like temperature, soil moisture, humidity, nitrate, potassium and phosphorus content of the soil. (Nikesh & Kawitkar, 2016). The sensors used in the system are explained below.

Temperature Sensor

NTC thermistor is used as a temperature sensor. NTC stands for Negative Temperature Coefficient. The sensor provides an analog output that can be converted into the ambient temperature in any unit require.

Soil Moisture Sensor

Soil moisture sensors measure the volumetric water content in soil.

Humidity Sensor

To measure humidity, amount of water molecules dissolved in the air of environment, a smart humidity sensor is used.

NPK(Nitrate, Potassium and phosphors) Sensor

Nitrate, Potassium and phosphors are the important macronutrients in the soil. With the help of this sensor will be able to get the nutrient level of a particular soil.

A wireless sensor network will be connect using Zigbee then the data will be sent to the microcontroller for later processing. The data collected by the sensors is send to the cloud for the processing of the data.

2. Data Maintenance System

The collected data must be properly maintained for the successful implementation of the proposed work so that it will be useful for the farmers to organize and process the data. With the help of the commercial software packages the maps can be produced for the historical data that allow the farmers to analyze the differences within fields. The proposed data maintenance system can work as the consultants to assist the farmers for taking decisions.

3. Data Analysis and Mining

Keeping historical data is meaningless unless it is properly analyzed and the results applied to solve a problem. Hence, this component is a very important in the proposed work that develops the agricultural plan. To achieve the desired outcome, the proposed work analyzes the collected data and applies the mining algorithms for making recommendations.

Role of Data Mining

Data mining plays a crucial role for decision making on several issues related to agriculture field. Data mining can be used for predicting future requirements and scope of agricultural processes. Data mining is the process that results in the discovery of new patterns in large data sets. Predictive data mining technique is used to predict future crop, weather forecasting, pesticides and fertilizers to be used, revenue to be generated and so on. The prediction problem can be solved by employing Data Mining techniques such as K Means, K nearest neighbour (KNN), Artificial Neural Network and support vector machine (SVM).

Data Mining Techniques

The following data mining techniques can be used for the proposed work depends upon the need.

1. Association Rule Mining
    a. It's to search unseen or desired crop patterns among the vast amount of data.
    b. Given a set of transactions, where each transaction is a set of literals, an association rule is a phrase of the form X => Y, where X and Y are sets of objects. (Transactions of the database which contain X tend to contain Y)
    c. Association rule algorithms like Apriori Algorithm(AA), Partition, Dynamic Hashing and Pruning(DHP), Dynamic Itemset Counting(DIC) and FP Growth(FPG) can be applied.

2. Classification
    a. It's used to extract models describing important data classes or to predict future data requirements like crop yield, disease, climate changes and soil conditions etc.
    b. It is a process in which a model learns to predict a class label from a set of training data which can then be used to predict discrete class labels on new samples.
    c. Learning approaches: semi-supervised learning, supervised learning and unsupervised learning can be implemented.
    d. The different classification techniques for discovering knowledge are Rule Based Classifiers, Bayesian Networks(BN), Decision Tree (DT),Nearest Neighbour(NN), Artificial Neural Network(ANN), Support Vector Machine (SVM), Rough Sets, Fuzzy Logic and Genetic Algorithms.
3. Clustering
    a. Clustering helps find new agricultural dataset by groupings the data instances into subsets in such a manner that similar instances are assembled together, while dissimilar instances belong to diverse groups.
    b. Clustering techniques like Hierarchical Methods(HM), Partitioning Methods (PM), Density-based Methods(DBM), Model-based Clustering Methods(MBCM), Grid-based Methods and Soft-computing Methods [fuzzy, neural network based], and Squared Error—Based Clustering (Vector Quantization).
4. Regression
    a. It is a process of learning a function that maps a data item to a real-valued prediction variable.
    b. Regression is used in predicting the nature of soil in a land and estimating the probability of crop support.
    c. Some techniques that can be used are Nonlinear Regression(NLR) and Linear Regression (LR).

4. Crop Monitoring Through Smart Irrigation

The farmers can choose the suitable type of crops for their land based on the given recommendation. The proposed system uses different emerging technologies to monitor the growth of crops. The variable nature of the fields may result in different amounts of water being needed in different areas.

Table 1.

S. No	Data Mining Techniques	Application
1.	Neural Networks	• Prediction of rainfall and weather forecast
2.	K-means	• Soil classification using GPS information • Disease analysis and prediction
3.	Fuzzy set	• Agricultural yield Prediction and weeds detection
4.	Decision Tree Analysis	• Analysis of some major crops like maize • Soil- site data in diverse land types
5.	Particle Swarm Optimization	• Agricultural water consumption forecasting
6.	Genetic Algorithm	• Plant disease detection

To automate the irrigation process, in order to provide an efficient usage of available water resources and to help farmers to be less concern about the irrigation process and irrigate the field without human supervision via droplet irrigation.

5. Image Analysis Collected Through Drones for Crop Disease Identification

In the field of agriculture, detection of disease in plants plays an important role. In plants, some general diseases seen are brown and yellow spots, early and late scorch, and others are fungal, viral and bacterial diseases. The existing method for plant disease detection is simply by naked eye. Plant disease identification by visual way is more laborious task and at the same time, less accurate and can be done only in limited areas. Whereas if automatic detection technique is used, it will take less efforts, less time and become more accurate. Automatic detection of the diseases by just seeing the symptoms on the plants makes it easier for earlier detection as well as cheaper.

Image processing is used for measuring affected area of disease and to determine the difference in the color of the affected area. Diverse image processing techniques with artificial neural network (ANN) lead to automatic early and accurately detection of plant diseases. After capturing the images, enhancement of image quality and extraction of feature of interest can be obtained using adaptive filtering and segmentation techniques. An artificial intelligence technique called multi-layer neural network can be trained with back-propagation algorithm to predict disease symptoms with respect to different disease patterns.

6. Alert System for Climatical Disasters

The proposed system also provides the alert system to communicate warnings and coordinate preparation activities during climatically disasters. The SMS will be sent to the farmers about the climate forecasting and types of preparation activities to be taken. The alert will be based on using monitoring data, including temperature and rainfall values, and state-of-the art climate models. Analysis on the observations and model-based predictions can be used to predict climate anomalies one or two seasons ahead.

7. Evaluation and Revision

After each cropping season, evaluation and revision of the agriculture plan is needed. To learn more about the field, more data can be gathered for each crop like productivity, profit, effective usage of water, proper utilization of pesticides and etc. These more information allows the further refinement of the agriculture plan.

CONCLUSION

Thus we conclude that IoT, sensor networks enables us to interact with the real world objects that enables us to connect agriculture to the IoT. It enhances the production of agricultural products and it is a comprehensive system designed to achieve precision in agriculture. IoT applications helps the farmers to use smart technologies to increase the sustainability in productions.

REFERENCES

Gondchawar & Kawitkar. (2016). Smart Agriculture Using IoT and WSN Based Modern Technologies. *International Journal of Innovative Research in Computer and Communication Engineering, 4*(6).

Jawad, H. M., Nordin, R., Gharghan, S. K., Jawad, A. M., & Ismail, M. (2017). Energy-Efficient Wireless Sensor Networks for Precision Agriculture: A Review. *Sensors, 17*(8), 1781. doi:10.339017081781 PMID:28771214

Managave, Savale, Ambekar, & Sathe. (2016). Precision Agriculture using Internet of Things and Wireless sensor Networks. *International Journal of Advanced Research in Computer Engineering & Technology, 5*(4).

Savale, Managave, Ambekar, & Sathe. (2015). Internet of Things in Precision Agriculture using Wireless Sensor Networks. *International Journal of Advanced Engineering & Innovative Technology, 2*(3).

Chapter 7
Selecting Location for Agro–Based Industry Using ELECTRE III Method

Seema Gupta Bhol
KIIT University (Deemed), India

Jnyana Ranjan Mohanty
KIIT University (Deemed), India

Proshikshya Mukherjee
KIIT University (Deemed), India

Prasant Kumar Pattnaik
KIIT University (Deemed), India

ABSTRACT

The Indian economy is driven by its agricultural sector. Industries based on agricultural produce are important as they give a competitive market for the agricultural production. Mustard is one of the major cash crops selected for this chapter. Mustard oil is used as cooking medium as well as other purpose in Indian households. Selecting the best location for setting up a mustard mill can be considered as a multiple criteria decision-making problem (MCDM), and ELECTRE III method is used and explained in detail to rank different location options in increasing order of suitability.

DOI: 10.4018/978-1-5225-9004-0.ch007

INTRODUCTION

India is leading producer of mustard; it is ranked third after China and Canada (Kumar, Premi, and Thomas). This plant belongs to cabbage family (Brassica); its botanical name is Brassica juncea. India produces 12 percent of world production (Kumar, Premi, and Thomas). India is also leading consumer of mustard. Mustard is second most important oil seed crop of India after Groundnut. The plant thrives in north and west India, mainly Satluj-Ganga plain. Indian mustard is cultivated in the states of Rajasthan, Uttar Pradesh, Haryana, Madhya Pradesh, and Gujarat which contribute 81.5% area and 87.5% production (Kumar, Premi, and Thomas). Alwar, Mathura, Morena are leading mustard producing districts in state of Rajasthan, Uttar Pradesh, Madhya Pradesh respectively.Rapeseed-Mustard is the main oilseed crop for the Rabi season which is planted on more than 74% area covered under oilseeds ("Executive Summary"). Mustard oil is main product of mustard seed. Oil can be extracted using cold pressing (kachhi ghani) or mechanical expelling and solvent extraction (Swati and Sehwag, 2015). Mustard oil accounts for 18% of Indian edible oil consumption ("Executive Summary"). Mustard seeds have 25-45% oil content and its oil cake makes important cattle feed and manure.

Taking into consideration the huge amount of mustard production and mustard oil consumption, it's profitable to setup mustard oil mill. The setting up of mustard mill depends upon various factors, thus the selection of suitable location can be considered as Multi criteria decision making problem. Selecting a location for the projects is common but difficult task. It is complicated because there are many criteria that needed to be compared. ELECTRE III method has several unique features like the concepts of outranking and the use of indifference and preference thresholds. The imprecise and uncertain nature of decision making can be incorporated, by using thresholds of indifference and preference. It allows the evaluation of alternatives to be undertaken as objectively as possible. Moreover, It is non-compensatory i.e. a very bad score on a criterion cannot be compensated by good scores on other criteria. ELECTRE models allow for incomparability and defines clear distinction between. Incomparability, and indifference (for alternatives a and b, when there is no clear evidence in favor of either a or b).

MULTI CRITERIA DECISION MAKING APPROACH

Multiple-criteria decision-making (MCDM) or multiple-criteria decision analysis (MCDA) is a sub-discipline of operations research. MCDM is concerned with structuring and solving decision and planning problems involving multiple criteria. Latest trends show application of MCDM method many agriculture related areas. Topsis and Fuzzy AHP are used for Efficient and effective manner of agriculture supply mechanization services to the farmer (Zangeneh, 2015). ELECTRE III is successfully implemented for ranking the agro-energy regions according to their potentials of biomass production (Preethi and Chandrasekar, 2015). ELECTRE applied in effective agriculture banking for providing financial sources to agricultural producers (Dincer, Hacioglu, and Yuksel, 2016). An agricultural product recommendation service on a mobile platform to assist consumers in deciding which agricultural product to buy (Macary et. al, 2013). ELECTRE acted as useful decision aid tool for implementing public agricultural and environmental policies for protecting the ecological areas by identifying erosion risk zones (Majdi, 2013). ELECTRE model applied for assessment of de-desertification alternatives those can be effective in controlling the reclamation of disturbed land and avoiding destruction of areas at risk (Solecka, 2014). Multiple criteria evaluation of 40 characteristics of e-learning platforms is done with ELECTRE (Kumar, Premi, and Thomas).

A MCDM can be represented by using a two-dimensional matrix known as performance matrix

Performance matrix describes relationship between available alternatives and critera for deciosn making. The Set of alternatives, A= {A_1, A_2, A_3.... A_n} forms rows and Set of Criteria C= {C_1,C_2,C_3,..,C_n} forms columns in the

Table 1. Performance matrix

		Criterion			
		C_1	C_2	.....	C_n
Alternatives	a_1	v_{11}	v_{12}	.....	v_{1n}
	a_2	v_{21}	v_{22}	.....	v_{2n}
	.	.	.	.	.
			.	.	.
	a_m	v_{m1}		.....	v_{mn}

performance matrix. Each cell, V_{ij} denotes the decision maker's preference for alternative i with respect to criterion j.

ELECTRE (Elimination and Choice Translating Algorithm)

The ELECTRE (Elimination and Choice Translating algorithm) was introduced by Benayoun, Roy and Sussman in 1968. The method was later developed by Bernard Roy (Roy, 1996). There are different versions of ELECTRE (I, II, III, IV and TRI), All methods share same conceptual foundation but differ both operationally and according to the type of decision problem.

In the present chapter, we will be focusing on ELECTRE III. This method is used when it is possible and desirable to quantify the relative importance of criteria (Buchanan, Sheppard, Vander Pooten, 1999). ELECTRE III is widely used multi-criteria decision-making tool as it can efficiently handle inaccurate, imprecise, uncertain data. ELECTRE III method is widely used in multi-criteria decision making in various areas. Supplier selection (Cristea and Cristea, 2017)(Galinska and Bielecki, 2017), partnership decision (Guarmieri and Hatakeyama, 2012), Green manufacturing (Arvind and Neeru, 2018), sustainability assessment (Majdi, 2013), Seamless Handoff in Mobile communication technology (Preethi and Chandrasekar, 2015), Infrastructure Investment (Rogers, Bruen, and Maystre, 2013), Business Considerations in Software prioritization (Sahaaya, Arul, and Suganya, 2016), integration of public transport (Solecka, 2014).

Electra III Method

Given a decision making problem with set of available alternatives and criteria . Also, for every criterion, decision maker must define the following:

1. Preference threshold (p)
2. Indifference threshold (q)
3. Veto thresholds (v)
4. Importance rating (wj) for each criterion j.

We will discuss them one by one

Let g_{ij}=1, 2..., n are defined criteria, and A is set of alternatives. The following three relations hold for two alternatives (a, b) A

- a P b (a is preferred to b) g(a) > g(b)
- a I b (a is indifferent to b) g(a) = g(b)
- a R b (a cannot be compared to b)

Indifference Threshold, q

It denotes the largest difference in performance between actions, that is judged as compatible with an indifference situation (Azziz, 2015). The preference relations are described as:

a **P** b (a is preferred to b) g(a) > g(b) + q

a **I** b (a is indifferent to b) |g(a) - g(b)| ≤q.

Preference Threshold, p

Buffer zone is introduced between indifference and strict preference, describing weak preference (Buchanan, Sheppard, and Vander Pooten, 1999). Preference threshold, p denotes the smallest difference in performance between actions, where a strict preference regarding the action with the best preference is expressed by the decision-maker (Kahraman, 2008). The preference relations are described as:

a P b (a is strongly preferred to b) g (a)-g (b) > p

a Q b (a is weakly preferred to b) q < g (a) - g (b) ≤p

a I b (a is indifferent to b; and b to a) |g (a) - g (b)| ≤q

Veto Threshold, v

The veto threshold, v_j, allows for the possibility of a **S** b to be refused totally if, for any one criterion j (Azziz, 2015). Veto threshold, v corresponds to the smallest difference in the performances between actions, where the decision-maker is incapable to assign one of the two situations (preference / indifference) (Galinska and Bielecki, 2017).

$g_j (b) > g_j (a) + v_j$.

Weights

Weights are "coefficients of importance", defined as set W:

$W = \{w_1, w_1, w_3, \ldots w_n\}$, where w_j is weight of criteria C_j

Steps in ELECTRE 3

The Eletre3 algorithm can be considered of having two main phases:

Phase1: Computation of Credibility matrix
Phase2: The ranking algorithm.

We will explain phase1 (Figure 1), followed by phase 2(Figure 2)

Figure 1. Phase 1 of ELECTRE III method

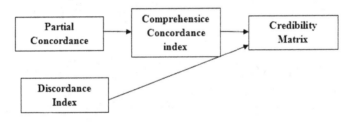

Figure 2. Phase 2 of ELECTRE III method

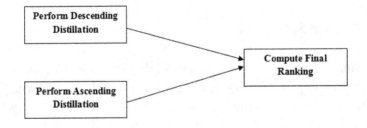

Phase 1: Construction of Credibility Matrix

To start with the process, decision maker defines

- Performance matrix
- Three thresholds (p, q, v) for each of the criterion
- Weights for each criterion

Once we have above mentioned values, we start with phase1 of the algorithm. Phase 1 is consisting of four main steps

Step1: Formation of Partial concordance matrix, corresponding to each criterion

Step2: Computation of comprehensive Concordance matrix

Step3: Formation of discordance matrix

Step4: Constructing outranking relation (Credibility matrix)

Concordance Matrix

The concordance matrix C (a, b), is a measurement to see to what extent *"action a outranks action b"* on criterion (Preethi and Chandrasekar, 2015). . A *concordance* principle requires that a majority of criteria, after considering their relative importance, is in favor of the assertion (Buchanan, Sheppard, and Vander Pooten, 1999). Let w_j be the importance coefficient or weight for criterion j. Concordance index between alternative a and b is defined as:

$$C(a,b) = \frac{1}{W} \sum_{j:gj(a) \geq gj(b)} wjcj(a,b) \tag{1}$$

Where

$$W = \sum_{j=1}^{r} wj$$

$$C_j(a,b) = \begin{cases} 1 & gj\left(a\right) + qj \geq gj\left(b\right) \\ 0 & gj\left(a\right) + pj \leq gj\left(b\right), \text{j} = 1,\dots, \text{n} \\ \dfrac{pj + gj\left(a\right) - gj\left(b\right)}{pj - qj} & otherwise \end{cases} \qquad (2)$$

Discordance Matrix

Discordance is the measure of the performances of the actions *a* and *b* oppose to the assertion *"action a is at least as good as action b"*. Veto threshold is used to compute discordance, the veto threshold, v_j, allows for the possibility of aSb to be rejected if:

for any one criterion j, g_j (b) > g_j (a) +v_j.

The discordance index for each criterion j, d_j (a, b) is calculated as:

$$d_j(a,b) = \begin{cases} 0, & if\ gj\left(a\right) + pj \geq gj\left(b\right) \\ 1, & if\ gj\left(a\right) + vj \leq g\left(b\right), j1,\dots, n \\ \dfrac{gj\left(b\right) - gj\left(a\right) - pj}{vj - pj}, & otherwise \end{cases} \qquad (3)$$

The Fuzzy Outranking Relation

In ELECTRE III, fuzzy outranking relation is defined by credibility matrix. A credibility matrix assesses the strength of the assertion that "a is at least as good as b" (Azziz, 2015) the Credibility of outranking is equal to concordance index where no criterion is discordant. However, when discordance exit, the concordance index is lowered in the relation to the importance of the discordance (Sahaaya, Arul, and Suganya, 2016). Credibility matrix is computed with the help of concordance and discordance matrix. The credibility degree for each pair (a, b) in A is defined as:

$$S(a,b) = \begin{cases} C(a,b) & \text{if } d(a,b) < C(a,b) \, \forall j \\ C(a,b) \prod\limits_{j \in J(a,b)} \dfrac{1 - dj(a)}{1 - C(a,b)} \end{cases} \qquad (4)$$

where $J(a,b)$ is the set of criteria Such that d(a,b) > C(a, b)

Thus, if the strength of the concordance is more than strength of discordance, then the concordance value remains unchanged. Otherwise, it is changes as per above equation. If the discordance is 1 for any pair of alternatives a,b A and any criterion j, then we show no confidence that aSb. Thus S(a,b) =0.

Phase 2: The Ranking Algorithm (Distillation)

After computation of fuzzy out ranking relation S, We start with phase 2 following the below mentioned steps:

Step1: Computation of Crisp out ranking relation T
Step2: Determining first complete preorderZ1 (Descending distillation)
Step3: Determining Second complete preorderZ2 (Ascending distillation)
Step4: Computation of final ranking Z using two complete preorders

Descending Distillation

In descending distillation process is used to computer first preorder Z1. It starts with selecting best alternative in the beginning and ends with worst alternative.

Ascending Distillation

The second pre-order Z2 is computed by ascending distillation. It starts with selecting worst alternative in the beginning and ends with best alternative.

Select the largest values in the credibility matrix and call it λ_0.

$\lambda_0 = \max S (a, b)$

The distillation discrimination threshold function, S (λ_0) is defined as follows:

$$S\ (\lambda_0) = \alpha + \beta\lambda \tag{5}$$

where, $S\ (\lambda_0)$ is the discrimination threshold at the maximum level of outranking λ. The values of α and β are usually 0.3 and -0.15.

After the computation of the cut off levels and determination of the distillation threshold, a crispy Outranking relation is obtained (Azziz, 2015), which is defined as:

$$T\ (a, b) = \begin{cases} 1, & if\ S\left(a,b\right) > \lambda - S\left(\lambda\right) \\ 0, otherwise \end{cases} \tag{6}$$

Next, qualification Q (a) is computed for each alternative a, as the number of alternatives that are Outranked by Alternative a minus the number of alternatives which outrank Alternative a. Q (a) is simply the row sum minus the column sum of the matrix T (Buchanan, Sheppard, and Vander Pooten, 1999). The set of alternatives having the largest qualification is the first distillate of D1. If D1 contains only one alternative, then previous procedure is repeated with remaining alternatives of A. Otherwise same procedure is executed within D1. The procedure is then repeated with remaining elements of A. The outcome of descending distillation is the first preorder Z1. The ascending distillation is carried out in a similar manner except that the alternatives with the smallest (instead of largest) qualification are retained first. The final partial pre-order is obtained by the intersection of the two complete pre-orders constructed in Descending and ascending phase (Azziz, 2015).

APPLICATION OF IN MUSTARD OIL MILL CASE

A Company wants to setup an oil mill to extract mustard oil. Company has option of setting up oil mill in either of three locations. The three cities under consideration are Alwar, Mathura, and Morena. The decision for the suitability of location depends upon three criteria. The different criteria of selecting location for oil mill includes availability of raw material i.e. mustard seeds, ease of doing business in that state, train and road connectivity with other places.

This is a multiple criteria decision problem. The objective is to maximize the profit. We will use ELECTRE III method to rank the alternatives from best to worst. A decision problem has two important components; a set of alternatives and a set of criteria.

Alternatives are three cities

- Alwar
- Mathura
- Morena

The three criteria are

- Raw material
- Ease of doing business
- Connectivity

The preference matrix relates specifically to the criteria and their relative importance.

The present case study compares four alternatives(Alwar, Mathura, Morena) based on three criteria namely, Raw material, Ease of doing business, connectivity. The performance matrix for three alternative locations and three criteria is described in Table 2.

The values for criteria are scaled from 1 to 10, 1 is least preferred and 10 is most preferred. We have also assigned weights to each of the criteria (one can compute them using AHP). The Decision maker assigned maximum importance to availability of raw material i.e.0.5. Ease of doing business is second important criteria after raw material and assigned weight (0.3). Connectivity ranks last in criteria importance with weight (0.2). Table 3 shows preference direction for criteria and associated weights

Table 2. Performance matrix

City	Ease of Business	Raw Material	Connectivity
Alwar	7	9	6
Mathura	8	6	7
Morena	5	7	6

Table 3. Criteria weights

Criteria	Direction of Preference	Weight
Ease of business	increasing	0.3
Raw material	increasing	0.5
Connectivity	increasing	0.2

We also define three thresholds as follows:

- Preference threshold, p=1
- Indifference threshold, q=0.5
- Veto threshold, v=2

Step 1: Computing Concordance index

The algorithm starts by computing the concordance index for each alternative over the Other alternatives and for each one of the criteria under consideration. We compute concordanceindex between pair of attributes for each criterion using eq 2.

The concordance calculations for Criteria 1 are:

$c1 (a_1, a2) = 0$, since $7+1 \leq 8$

$c1 (a_1, a3) = 1$, since $7 + 0.5 \geq 7$

$c1 (a_2, a_1) = 1$, since $8 + 0.5 \geq 7$

$c1 (a_2, a_3) = 1$, since $8 + 0.5 \geq 5$

$c1 (a_3, a_1) = 0$, since $5 + 1 \leq 7$

$c1 (a_3, a_2) = 0$, since $5 + 1 \leq 8$

Table 4. Partial concordance matrix for criterion 1 i.e. ease of doing business

C1 (...,)	Alwar	Mathura	Morena
Alwar	1	0	1
Mathura	1	1	1
Morena	0	0	1

The concordance calculations for Criteria 2 are:

$c_2 (a_1, a2) = 1$, since $9 + 0.5 \geq 6$

$c_2 (a_1, a3) = 1$, since $9 + 0.5 \geq 7$

$c_2 (a_2, a_1) = 0$, since $6 + 1 \leq 9$

$c_2 (a_2, a_3) = 0$, since $6 + 1 \leq 7$

$c_2 (a_3, a_1) = 0$, since $7 + 1 \leq 9$

$c_2 (a_3, a_2) = 1$, since $7 + 0.5 \geq 6$

The concordance calculations for Criteria 3 are:

$c3 (a_1, a2) = 0$, since $6 + 1 \leq 7$

$c3 (a_1, a3) = 1$, since $6 + 0.5 \geq 6$

$c3 (a_2, a_1) = 1$, since $7 + 0.5 \geq 6$

$c3 (a_2, a_3) = 1$, since $7 + 0.5 \geq 6$

Table 5. Partial concordance matrix for criterion2 i.e. raw material

C2 (..,..)	Alwar	Mathura	Morena
Alwar	1	1	1
Mathura	0	1	0
Morena	0	1	1

Table 6. Partial concordance matrix for criteria3 i.e. connectivity

C3 (..,..)	Alwar	Mathura	Morena
Alwar	1	0	1
Mathura	1	1	1
Morena	1	0	1

c3 (a_3, a_1) = 1, since $6 + 0.5 \geq 6$

c3 (a_3, a_2) = 0, since $6 + 1 \leq 7$

Now we calculate Comprehensive concordance index using eq1 with the help of Table 2, Table 4, Table 5 and Table 6.

The concordance index between same alternatives is always 1. Thus:

C $(a_1, a1)$ =1

C $(a_2, a2)$ =1

C $(a_3, a3)$ =1

The Concordance calculations for comparison of an alternative with other alternatives are as follows:

$$\textbf{C} \ (a_1, a_2) = \frac{0X0.3 + 1X0.5 + 0X0.2}{0.3 + 0.5 + 0.2} = 0.5$$

$$\textbf{C} \ (a_1, a_3) = \frac{1X0.3 + 1X0.5 + 1X0.2}{0.3 + 0.5 + 0.2} = 1.0$$

$$\textbf{C} \ (a_2, a_1) = \frac{1X0.3 + 0X0.5 + 1X0.2}{0.3 + 0.5 + 0.2} = 0.5$$

$$\textbf{C} \ (a_2, a_3) = \frac{1X0.3 + 0X0.5 + 1X0.2}{0.3 + 0.5 + 0.2} = 0.5$$

$$\textbf{C} \ (a_3, a_1) = \frac{0X0.3 + 0X0.5 + 1X0.2}{0.3 + 0.5 + 0.2} = 0.2$$

$$\textbf{C} \ (a_3, a_2) = \frac{1X0.3 + 1X0.5 + 1X0.2}{0.3 + 0.5 + 0.2} = 1.0$$

Table 7. Comprehensive concordance matrix

C (...,)	Alwar	Mathura	Morena
Alwar	1	0.5	1
Mathura	0.5	1	0.5
Morena	0.2	1	1

Step 2: Computing Discordance Index

Now, we will compute the discordance index for each alternative over the other alternatives and for each one of the criteria under consideration. For each pair of alternatives, discordance index is computed for each of the criterions using eq3.

The discordance calculations for Criteria 1 are:

$d1\ (a_1, a2) = 0$, since $7+1 \geq 8$

$d1\ (a_1, a3) = 0$, since $7 + 1 \geq 5$

$d1\ (a_2, a_1) = 0$, since $8 + 1 \geq 7$

$d1\ (a_2, a_3) = 0$, since $8 + 1 \geq 5$

$d1\ (a_3, a_1) = 1$, since $5 + 2 \leq 7$

$d1\ (a_3, a_2) = 1$, since $5 + 2 \leq 8$

The discordance calculations for Criteria 2 are:

$d_2\ (a_1, a_2) = 0$, since $9 + 1 \geq 6$

Table 8. Discordance matrix for criterion 1, i.e. ease of doing business

	Alwar	Mathura	Morena
Alwar	0	0	0
Mathura	0	0	0
Morena	1	1	0

$d_2 (a_1, a_3) = 0$, since $9 + 1 \geq 7$

$d_2 (a_2, a_1) = 1$, since $6 + 2 \leq 9$

$d_2 (a_2, a_3) = 0$, since $6 + 1 \geq 7$

$d_2 (a_3, a_1) = 1$, since $7 + 2 \leq 9$

$d_2 (a_3, a_2) = 0$, since $7 + 1 \geq 6$

The discordance calculations for Criteria 3 are:

$d3 (a_1, a_2) = 0$, since $6 + 1 \geq 7$

$d3 (a_1, a_3) = 0$, since $6 + 1 \geq 6$

$d3 (a_2, a_1) = 0$, since $7 + 1 \geq 6$

$d3 (a_2, a_3) = 0$, since $7 + 1 \geq 6$

$d3 (a_3, a_1) = 0$, since $6 + 1 \geq 6$

$d3 (a_3, a_2) = 0$, since $6 + 1 \geq 7$

Table 9. Discordance matrix for criterion 2 i.e. raw material

	Alwar	Mathura	Morena
Alwar	0	0	0
Mathura	1	0	0
Morena	1	0	0

Table 10. Discordance matrix for criterion 3, i.e. connectivity

	Alwar	Mathura	Morena
Alwar	0	0	0
Mathura	0	0	0
Morena	0	0	0

Step 3: The Fuzzy Outranking Relation, Credibility Matrix, S

Credibility index, S will be calculated using eq 4. The result of computations is stored in Table 11.

Step 4: The Ranking Algorithm

In this step the two complete pre-orders will be computed, the credibility matrix (Table 11) will be used to carryout Descending and ascending distillations.

Descending distillation will be carried out using eq 6. Largest score in credibility matrix is selected and set as λ_0. Thus $\lambda_0=1$. Now we will calculate discrimination threshold $S(\lambda_0)$.

Now, we put $\lambda_0=1$, $\alpha=0.3$ and, $\beta=-0.15$ in eq 5, we get

$$S(\lambda_0) = 0.3 + (-0.15 \times 1)$$

$$=1.5$$

Cut-off level for outranking, λ_1 will be computed next

$$\lambda_1 = \lambda_0 - S(\lambda_0)$$

$$= 1 - 1.5 = 0.85$$

In credibility matrix S, 0.50 is the richest credibility degree lower than 0.85.

To compute Matrix T, We put 1 for all the values greater than equal to 5.0 and 0 otherwise. Matrix T is shown in the Table 12.

Qualification for each alternative will be calculated from strength and weakness of the alternative

Table 11. Credibility matrix

S	Alwar	Mathura	Morena
Alwar	1	0.5	1
Mathura	0	1	0.5
Morena	0	0	1

Table 12. First distillation, crispy outranking matrix

T1	Alwar	Mathura	Morena
Alwar	1	1	1
Mathura	0	1	1
Morena	0	0	1

Q (a) = Strength (a) - Weakness (a)

Qualification, Q (a), for all an A is the number of alternatives that are Outranked by a minus the number of alternatives which outrank alternative a. Thus Q (a) Row sum minus the column sum of the matrix T. Qualification for all the alternatives are presented in Table 13.

The Qualification of alternative Alwar is highest. The first distillate of distillation process D1 is {Alwar}. The same procedure will be repeated A \ D1 (where A is set of alternatives).

Now, we will carry distillation process on {Mathura, Morena}.

We repeat the same procedure to compute λ_0, S (λ_0).

$\lambda_0 = 1$

S (λ_0) = 0.3 + (-0.15 x1) =1.5

Table 13. Qualifications (Distillation 1)

D1	Strength	weakness	Qualification
Alwar	3	1	2
Mathura	2	2	0
Morena	1	3	-2

Table 14. Credibility matrix (distillation2)

S	Mathura	Morena
Mathura	1	0.5
Morena	0	1

$\lambda_2 = \lambda_{0-} S (\lambda_0)$

$= 1 - 1.5 = 0.85$

In credibility matrix for second distillation (Table 14), 0.50 is the richest credibility degree lower than 0.85. Following equation 6, the crispy outranking matrix is defined as shown in Table 15.

Now, Qualifications for each of the alternatives are compute and presented in Table 16.

The Qualification of alternative Mathura is highest. The second distillate D2 is {Mathura}.

Thus, the first preorder is Alwar, Mathura, and Morena

Similarly, Descending distillation can be performed. The process is same as descending distillation; the only difference is, instead of selecting alternative with best ranking, alternative with worst ranking will be selected.

The Final preorder is shown in Figure 3.

From Figure 3, we conclude that Alwar is best option for setting up mustard oil mill, following the mentioned criteria under consideration. The Mathura ranked second and Morena third.

RESULT ANALYSIS AND DISCUSSIONS

For effective management of agricultural resources and decision-making, multi criteria analysis is widely used in agriculture sector by modelling

Table 15. Second distillation, crispy outranking matrix

T2	Mathura	Morena
Mathura	1	1
Morena	0	1

Table 16. Qualifications (Distillation2)

D2	Strength	Weakness	Qualification
Mathura	2	1	1
Morena	1	2	-1

Figure 3. Final preorder

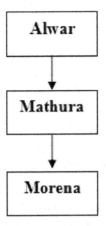

agricultural decision-making into mathematical model. ELECTRE III method has two main phases. In phase 1, Firstly the Concordance index is calculated and Discordance index is measured. Finally, Credibility index is computed. In Phase 2 Outranking Relationships among the different alternatives is computed (see Table 12). From that, Weakness and Strength for each alternative is computed. Comparative ranking of alternatives on the basis of ELECTRE III allots the Location ALWAR highest rank followed by Morena and Matura. Thus Alwar is the best choice for mustard oil mill following the above mentioned criteria. However, it may be mentioned that the ranking depends on the judgements of relative importance made by the Decision maker (Arvind and Neeru, 2018). The ranking may change if the user assigns different weights to the criteria.

CONCLUSION

This chapter describes application of the ELECTRE III decision tool to a site selection for setting up mustard oil mill. The case study illustrates how the ELECTRE III model takes complex information and uses it to rank the various project alternatives considered. A weighting system is used for estimating the relative importance of criteria. The detailed process of how the ELECTRE III works to compute ranking of various alternatives is described in the chapter. The calculation steps are elaborated in a case study. ELECTRE III methods make use of thresholds to deal important aspect of the real world, We take appropriate threshold values suitable for the pilot data . By Calculating

Concordance, Discordance and Credibility matrices we can select the best location for setting up the mustard oil mill. On the other hand, the ELECTRE method has some drawbacks. The major weakness is that if the parameters are not chosen rightly the method would result in an inefficient usage of resources performance (Preethi and Chandrasekar, 2015). Thus by choosing correct parameters these methods provide a useful tool for comparing various locations available for setting up mustard oil mill.

The chapter has methodological importance as it indicates how the complex analyses and evaluation of the location alternatives should be carried out to construct the final ranking of the alternatives from the best alternative to the worst.The method is validated using a case study with fewer requirements. Literature shows that the method is scalable, but the validation is not done in this research work. In the future, the method is planned to be validated with large scale requirements (Sahaaya, Arul, and Suganya, 2016).

REFERENCES

Azziz. (2015). *Multiple Criteria Outranking Algorithm: Implementation and Computational Tests* (M.Sc thesis).

Buchanan, Sheppard, & Vander Pooten. (1999). *Project ranking using electre iii*. Academic Press.

Cristea & Cristea. (2017). A multi-criteria decision-making approach for Supplier selection in the flexible packaging Industry. *MATEC Web of Conferences*, 94.

Dincer, Hacıoğlu, & Yüksel. (2016). Managerial and Market-Based Appraisal of Agriculture Banking Using ANP and ELECTRE Method. *Management and Organizational Studies, 3*(3).

Galinska & Bielecki. (2017). *Multiple criteria evaluation of suppliers in company operating in clothing industry*. 17th international scientific conference Business Logistics in modern management, Osijek, Croatia.

Guarnieri & Hatakeyama. (2012). The process of decision-making regarding partnerships including a multicriteria method. *International Journal of Advanced Manufacturing Systems*. Retrieved from https://www.researchgate.net/publication/235982266

Jayant & Chaudhary. (2018). *A decision-making framework model of cutting fluid selection for Green manufacturing: a synthesis of 3 mcdm approaches.* Retrieved from https://www.researchgate.net/publication/323748301

Jiunn-Woei, L., & Ke, C.-K. (2016, November). Using a modified ELECTRE method for an agricultural product recommendation service on a mobile device. *Computers & Electrical Engineering,* 277–288.

Kahraman, C. (2008). *Fuzzy multi-criteria decision-making theory and Applications with Recent developments.* Academic Press.

Kirkenidis, Andreopoulou, & Manos. (2016). *Evaluation of e-learning platforms suitable for Agriculture and Forestry Higher Schools: A case study using ELECTRE III.* Academic Press.

Kumar, Premi, & Thomas. (n.d.). *Rapeseed-Mustard cultivation in India-An Overview.* National Research Centre on Rapeseed-Mustard, Bharatpur.

Macary, Dias, Figueira, & Roy. (2013). *A Multiple Criteria Decision Analysis Model Based on ELECTRE TRI-C for Erosion Risk Assessment in Agricultural Areas.* Academic Press.

Majdi. (2013). *Comparative evaluation of promethee and electre with application to sustainability assessment* (Thesis). Concordia University, Montreal, Canada.

Moulogianni & Bournaris. (2017). Biomass Production from Crops Residues: Ranking of Agro-Energy Regions. *Energies, 10*(1061). doi:10.3390/en10071061

Preethi, G.A., & Chandrasekar, C. (2015). Seamless Handoff using Electre III and Promethee Methods. *International Journal of Computer Applications, 126*(13).

Rogers, Bruen, & Maystre. (2013). *Electre and Decision Support: Methods and Applications in Engineering and Infrastructure Investment.* Springer Science & Business Media.

Sadeghiravesh, M.H., Zehtabian, G.R., & Khosravi, H. (2014). Application of AHP and ELECTRE models for assessment of de-desertification alternatives. *Desert, 19*(2), 141-153.

Sahaaya Arul Mary, S. A., & Suganya, G. (2016, May). Multi-Criteria Decision Making Using ELECTRE. *Circuits and Systems*, *7*(6), 1008–1020. doi:10.4236/cs.2016.76085

Solecka. (2014). Urban public transport system assessment; integration of public transport. *Multiple-Criteria Decision Making*.

Swati & Sehwag. (2015). A brief overview: Present status on utilization of mustard oil and cake. *Indian Journal of Traditional Knowledge*, *14*(2), 244–250.

The Solvent Extractors Association of India. (n.d.). *Executive Summary Rapeseed-Mustard Crop Survey 2016-17*. Author.

Zangeneh, Akram, Nielsen, & Keyhani. (2015). *Developing location indicators for Agricultural Service Center: A Delphi–TOPSIS–FAHP approach*. Academic Press.

Conclusion

In this book, Internet of Things (IoT) and Wireless Sensor Networks (WSNs) are related to agriculture technology, applications of WSN and IoT application in agriculture have been described. The main goal is to focus the development of effective data computing operations on agricultural advancement that are fully supported by IoT, cloud computing and WSN systems. The importance of IoT and WSN to agriculture technology have been also described. Some of the techniques in WSN and IoT in agriculture have been discussed in various chapters. In the introduction, the importance of agriculture technology and its advancements in WSNand IoT has been discussed.

Chapter 1 reviewed the of the various IoT techniques used in smart farming. In agricultural field, it is used to provide proper information about the crop, weather, properties of soil, availability of machinery, supply chain management, smart tagging system, etc. According to the economic condition of the farmer all these facilities may not be affordable for small and marginal farmers. So the IoT devices should be cost effective, and easily available to the farmer. Future work will consist of a smart and efficient system of security which would provide protection from outsiders in the field or any rodents or animals.

Chapter 2 discusses different applications of wireless sensor network in health care domain. Wireless Sensor Networks (WSNs) commonly knew as the Internet of Things (IoT) enables a global approach to the health care system infrastructure development. This leads to the e-health system that, in real time manner, supplies a valuable set of information relevant to all of the stakeholders regardless of their current location. Commercial systems in this area usually do not meet the general patient needs, and those that do are generally economically unacceptable due to the high operational and development costs. Here, the interoperability of sensor data to build promising and interoperable domain-specific or cross-domain IoT applications.

Chapter 3 discusses the energy efficient routing with mobile sink protocols that are more suitable to strengthen the agriculture. Here, classifying aforesaid protocols into three different categories viz. Hierarchical based, tree based and virtual- structure based routing. Here, packet congestion is minimized along with energy optimization among sensor nodes. Hence dynamic rearrangement of the routing tree makes the system more efficient towards network delay and improving of network lifetime.

Chapter 5 developed a power mapping localization algorithm based on the Monte Carlo method using a discrete antithetic approach called Antithetic Markov Chain Monte Carlo (AMCMC). Here, focused on solving two major problems in WSN, thereby increasing the accuracy of the localization algorithm and discrete power control. Consecutively, the work is focused to reduce the computational time, while finding the location of the sensor. The model achieves the power controlling strategy using discrete power levels (CC2420 radio chip) which allocate the power, based on the event of each sensor node. By utilizing this discrete power mapping method, all the high level parameters are considered for WSN. To improve the overall accuracy, the antithetic sampling is used to reduce the number of unwanted sampling, while identifying the sensor location in each transition state.

Chapter 5 discussed that multiple nodes are required for co-perative communication, the Low Energy Adaptive clustering Hierarchy (LEACH) and LEACH - Vector Quantization (LEACH-V) are used for cluster and active cluster head (CH) formation. Further Dijkstra Algorithm is used to find the shortest path between the active cluster heads (CHs) and high energy utilization respectively. The main issue of inter-cluster communication is carried out in earlier work using LEACH and LEACH-V protocols. The proposed work illustrates the LEACH-Vector Quantization Dijkstra (LEACH-VD) protocol, for shortest path active cluster head (CH) communication in a Cooperative communication network.In the application point of view, LEACH-VD performs the lowest energy path.LEACH-V provides the intra -cluster communication between the cluster head and using Dijkstra Algorithm, the minimum distance is calculated connecting the active cluster heads which creates the shortest path results in energy efficient technique.

Chapter 6 enhanced the productivity of rural Tamil Nadu in order to meet the growing demands of our country's food supply chain. Agriculture is a sector that holds great promise to Indian economic growth. Production in rural Tamil Nadu is extremely low due to unscientific farming practices. The major challenges faced in Tamil Nadu agriculture are crop mapping, yield prediction, quality of food produced, irrigation management, variable rate fertilizer and pesticide due to lack of technical knowledge. Precision Agriculture (PA) rules out all drawbacks of traditional agriculture.

Chapter 7 gives methodological importance as it indicates how the complex analyses and evaluation of the location alternatives should be carried out to construct the final ranking of the alternatives from the best alternative to the worst. The method is validated using a case study with fewer requirements. Literature shows that the method is scalable, but the validation is not done in this research work. Mustard being one of the major cash crop, selected for this chapter. Mustard oil is used as cooking medium as well as much other purpose in Indian households. Selecting the best location for Setting up of mustard mill can be considered as multiple criteria decision making, problem (MCDM) and ELECTRE III methods is used and explained in detail to rank different location options in increasing order of suitability.

IoT and WSNs are today widely increasing the applications of agriculture domain. Our goal in this edited book was to make available a comprehensive, responsive outline to IoT and WSN application of agriculture technology. We wish the chapters and the book express not only some significant information about agriculture research and also the keenness that authors share about promising and expeditious research field.

Proshikshya Mukherjee
KIIT University (Deemed), India

Prasant Kumar Pattnaik
KIIT University (Deemed), India

Surya Narayan Panda
Chitkara University, India

Related Readings

To continue IGI Global's long-standing tradition of advancing innovation through emerging research, please find below a compiled list of recommended IGI Global book chapters and journal articles in the areas of agricultural systems, wireless sensor networks, and the internet of things. These related readings will provide additional information and guidance to further enrich your knowledge and assist you with your own research

Agarwal, P., Singh, V., Saini, G. L., & Panwar, D. (2019). Sustainable Smart-Farming Framework: Smart Farming. In R. Poonia, X. Gao, L. Raja, S. Sharma, & S. Vyas (Eds.), *Smart Farming Technologies for Sustainable Agricultural Development* (pp. 147–173). Hershey, PA: IGI Global. doi:10.4018/978-1-5225-5909-2.ch007

Ahmedi, F., Ahmedi, L., O'Flynn, B., Kurti, A., Tahirsylaj, S., Bytyçi, E., ... Salihu, A. (2018). InWaterSense: An Intelligent Wireless Sensor Network for Monitoring Surface Water Quality to a River in Kosovo. In P. Papajorgji & F. Pinet (Eds.), *Innovations and Trends in Environmental and Agricultural Informatics* (pp. 58–85). Hershey, PA: IGI Global. doi:10.4018/978-1-5225-5978-8.ch003

Al Manir, M. S., Spencer, B., & Baker, C. J. (2018). Decision Support for Agricultural Consultants With Semantic Data Federation. *International Journal of Agricultural and Environmental Information Systems*, 9(3), 87–99. doi:10.4018/IJAEIS.2018070106

Arunachalam, M., & Arunachalam, M. (2019). Identification of Good One From the Damaged Crops/Fruits Using Decision-Level Information Matching. In N. Razmjooy & V. Estrela (Eds.), *Applications of Image Processing and Soft Computing Systems in Agriculture* (pp. 80–99). Hershey, PA: IGI Global. doi:10.4018/978-1-5225-8027-0.ch003

Asani, A. M., Lukman, S., & Oke, I. A. (2019). Application of Software in Soil and Groundwater Recharge Estimation in Ilorin, Nigeria. In N. Razmjooy & V. Estrela (Eds.), *Applications of Image Processing and Soft Computing Systems in Agriculture* (pp. 184–207). Hershey, PA: IGI Global. doi:10.4018/978-1-5225-8027-0.ch008

Atungulu, G., & Mohammadi-Shad, Z. (2019). Reference on Mycotoxins Occurrence, Prevalence, and Risk Assessment in Food Systems. In P. Gaspar & P. da Silva (Eds.), *Novel Technologies and Systems for Food Preservation* (pp. 294–343). Hershey, PA: IGI Global. doi:10.4018/978-1-5225-7894-9.ch012

Atungulu, G., & Shafiekhani, S. (2019). Reference on Rice Quality and Safety. In P. Gaspar & P. da Silva (Eds.), *Novel Technologies and Systems for Food Preservation* (pp. 226–274). Hershey, PA: IGI Global. doi:10.4018/978-1-5225-7894-9.ch010

Badea, M., & Presada, D. (2016). Developing Students' English Language Skills and Cultural Awareness by Means of Food Topics. In A. Jean-Vasile (Ed.), *Food Science, Production, and Engineering in Contemporary Economies* (pp. 176–202). Hershey, PA: IGI Global. doi:10.4018/978-1-5225-0341-5.ch007

Barakabitze, A. A., Fue, K. G., Kitindi, E. J., & Sanga, C. A. (2016). Developing a Framework for Next Generation Integrated Agro Food-Advisory Systems in Developing Countries. *International Journal of Information Communication Technologies and Human Development*, 8(4), 13–31. doi:10.4018/IJICTHD.2016100102

Bergander, M. J., & Kapayeva, S. D. (2019). Solar Refrigeration for Post-Harvest Storage of Agricultural Products. In P. Gaspar & P. da Silva (Eds.), *Novel Technologies and Systems for Food Preservation* (pp. 108–139). Hershey, PA: IGI Global. doi:10.4018/978-1-5225-7894-9.ch005

Bhalani, D. V., Chandel, A. K., & Thakur, P. S. (2019). Food Quality and Safety Regulation Systems at a Glance. In P. Gaspar & P. da Silva (Eds.), *Novel Technologies and Systems for Food Preservation* (pp. 275–293). Hershey, PA: IGI Global. doi:10.4018/978-1-5225-7894-9.ch011

Bharti, P. K. (2019). Soil Quality Assessment at Nganglam, Bhutan. In A. Rathoure (Ed.), *Amelioration Technology for Soil Sustainability* (pp. 179–194). Hershey, PA: IGI Global. doi:10.4018/978-1-5225-7940-3.ch011

Bhatnagar, V., & Poonia, R. C. (2019). Sustainable Development in Agriculture: Past and Present Scenario of Indian Agriculture. In R. Poonia, X. Gao, L. Raja, S. Sharma, & S. Vyas (Eds.), *Smart Farming Technologies for Sustainable Agricultural Development* (pp. 40–66). Hershey, PA: IGI Global. doi:10.4018/978-1-5225-5909-2.ch003

Budnikov, D., Vasiliev, A. N., Vasilyev, A. A., Morenko, K. S., Mohamed, I. S., & Belov, A. (2019). The Application of Electrophysical Effects in the Processing of Agricultural Materials. In V. Kharchenko & P. Vasant (Eds.), *Advanced Agro-Engineering Technologies for Rural Business Development* (pp. 1–27). Hershey, PA: IGI Global. doi:10.4018/978-1-5225-7573-3.ch001

Caliman, B. F., & Ene, C. (2016). Sensory Evaluation in Food Manufacturing: Practical Guidelines. In A. Jean-Vasile (Ed.), *Food Science, Production, and Engineering in Contemporary Economies* (pp. 294–314). Hershey, PA: IGI Global. doi:10.4018/978-1-5225-0341-5.ch012

Carli, G., Canavari, M., & Grandi, A. (2018). Introducing Activity-Based Costing in Farm Management: The Design of the FarmBO System. In P. Papajorgji & F. Pinet (Eds.), *Innovations and Trends in Environmental and Agricultural Informatics* (pp. 252–272). Hershey, PA: IGI Global. doi:10.4018/978-1-5225-5978-8.ch010

Carmenado, I. D., Hernandez, H. B., Mendez, M. R., & Ferrer, C. G. (2016). Managing for the Sustained Success of Organic Food Associations: A Sustainable Management Approach from "Working with People" Model. In A. Jean-Vasile (Ed.), *Food Science, Production, and Engineering in Contemporary Economies* (pp. 25–43). Hershey, PA: IGI Global. doi:10.4018/978-1-5225-0341-5.ch002

Chaudhry, S., & Garg, S. (2019). Smart Irrigation Techniques for Water Resource Management. In R. Poonia, X. Gao, L. Raja, S. Sharma, & S. Vyas (Eds.), *Smart Farming Technologies for Sustainable Agricultural Development* (pp. 196–219). Hershey, PA: IGI Global. doi:10.4018/978-1-5225-5909-2.ch009

Chowhan, R. S., & Dayya, P. (2019). Sustainable Smart Farming for Masses Using Modern Ways of Internet of Things (IoT) Into Agriculture. In J. Rodrigues, A. Gawanmeh, K. Saleem, & S. Parvin (Eds.), *Smart Devices, Applications, and Protocols for the IoT* (pp. 189–219). Hershey, PA: IGI Global. doi:10.4018/978-1-5225-7811-6.ch009

Cristina, G., & Iridon, C. (2016). Food Cultural Values: An Approach to Multiculturality and Interculturality. In A. Jean-Vasile (Ed.), *Food Science, Production, and Engineering in Contemporary Economies* (pp. 125–145). Hershey, PA: IGI Global. doi:10.4018/978-1-5225-0341-5.ch005

Drăgoi, M. C. (2016). Health Determinants: Nutrition-Related Facts. In A. Jean-Vasile (Ed.), *Food Science, Production, and Engineering in Contemporary Economies* (pp. 393–417). Hershey, PA: IGI Global. doi:10.4018/978-1-5225-0341-5.ch017

Drevin, V., Fomichev, V., Yudaev, I. V., Minchenko, L., Gizzatova, G., Kucherova, I., ... Rybintsev, I. (2019). Application of Solutions of the Electrochemical Processed Mineral "Bishofit" in Plant Production. In V. Kharchenko & P. Vasant (Eds.), *Advanced Agro-Engineering Technologies for Rural Business Development* (pp. 320–345). Hershey, PA: IGI Global. doi:10.4018/978-1-5225-7573-3.ch012

Dubrovin, A. (2019). Control of Advanced Fodder Disinfection in Terms of Economic Criteria. In V. Kharchenko & P. Vasant (Eds.), *Advanced Agro-Engineering Technologies for Rural Business Development* (pp. 431–439). Hershey, PA: IGI Global. doi:10.4018/978-1-5225-7573-3.ch016

Durugkar, S. R., & Poonia, R. C. (2019). Peregrinating Gardens From Traditional to Most Advanced Handy Approach for Avoiding the Unnecessary Utilization of Resources. In R. Poonia, X. Gao, L. Raja, S. Sharma, & S. Vyas (Eds.), *Smart Farming Technologies for Sustainable Agricultural Development* (pp. 174–195). Hershey, PA: IGI Global. doi:10.4018/978-1-5225-5909-2.ch008

Durugkar, S. R., Poonia, R. C., & Naik, R. B. (2017). Case Study on WSN Based Smart Home Garden with Priority Driven Approach. *Journal of Cases on Information Technology, 19*(4), 37–48. doi:10.4018/JCIT.2017100104

Edoh-Alove, E., Bimonte, S., Pinet, F., & Bédard, Y. (2018). New Design Approach to Handle Spatial Vagueness in Spatial OLAP Datacubes: Application to Agri-Environmental Data. In P. Papajorgji & F. Pinet (Eds.), *Innovations and Trends in Environmental and Agricultural Informatics* (pp. 129–155). Hershey, PA: IGI Global. doi:10.4018/978-1-5225-5978-8.ch006

Gallyamova, T. R., Vozmischev, A. L., Khokhriakov, N. V., Baltachev, V. G., & Ponomaryova, S. Y. (2019). On Efficiency Rising of the Technological LED Lighting for the Floor Poultry Keeping. In V. Kharchenko & P. Vasant (Eds.), *Advanced Agro-Engineering Technologies for Rural Business Development* (pp. 123–148). Hershey, PA: IGI Global. doi:10.4018/978-1-5225-7573-3.ch005

Gaspar, P. D., da Silva, P. D., Andrade, L. P., Nunes, J., & Santo, C. E. (2019). Technologies for Monitoring the Safety of Perishable Food Products. In P. Gaspar & P. da Silva (Eds.), *Novel Technologies and Systems for Food Preservation* (pp. 190–225). Hershey, PA: IGI Global. doi:10.4018/978-1-5225-7894-9.ch009

Gere, A., Radványi, D., Sciacca, R., & Moskowitz, H. (2018). Mental Informatics and Agricultural Issues: Global Change vs. Sustainable Agriculture. In P. Papajorgji & F. Pinet (Eds.), *Innovations and Trends in Environmental and Agricultural Informatics* (pp. 1–37). Hershey, PA: IGI Global. doi:10.4018/978-1-5225-5978-8.ch001

Gill, S. S., Chana, I., & Buyya, R. (2017). IoT Based Agriculture as a Cloud and Big Data Service: The Beginning of Digital India. *Journal of Organizational and End User Computing, 29*(4), 1–23. doi:10.4018/JOEUC.2017100101

Guma, I. P., Rwashana, A. S., & Oyo, B. (2018). Food Security Indicators for Subsistence Farmers Sustainability: A System Dynamics Approach. *International Journal of System Dynamics Applications, 7*(1), 45–64. doi:10.4018/IJSDA.2018010103

Jat, D. S., Limbo, A. S., & Singh, C. (2019). Internet of Things for Automation in Smart Agriculture: A Technical Review. In R. Poonia, X. Gao, L. Raja, S. Sharma, & S. Vyas (Eds.), *Smart Farming Technologies for Sustainable Agricultural Development* (pp. 93–105). Hershey, PA: IGI Global. doi:10.4018/978-1-5225-5909-2.ch005

Jat, D. S., & Madamombe, C. G. (2019). Wireless Sensor Networks Technologies and Applications for Smart Farming. In R. Poonia, X. Gao, L. Raja, S. Sharma, & S. Vyas (Eds.), *Smart Farming Technologies for Sustainable Agricultural Development* (pp. 25–39). Hershey, PA: IGI Global. doi:10.4018/978-1-5225-5909-2.ch002

Jean-Vasile, A., & Alecu, A. (2016). Trends and Transformations in European Agricultural Economy, Rural Communities and Food Sustainability in Context of New Common Agricultural Policy (CAP) Reforms. In A. Jean-Vasile (Ed.), *Food Science, Production, and Engineering in Contemporary Economies* (pp. 1–24). Hershey, PA: IGI Global. doi:10.4018/978-1-5225-0341-5.ch001

Kalaiselvam, S., Rajan, D., & Ali, I. H. (2019). Thermal Technologies and Systems for Food Preservation. In P. Gaspar & P. da Silva (Eds.), *Novel Technologies and Systems for Food Preservation* (pp. 140–159). Hershey, PA: IGI Global. doi:10.4018/978-1-5225-7894-9.ch006

Kapp, C. Jr, Caires, E. F., & Guimarães, A. M. (2018). Discriminating Biomass and Nitrogen Status in Wheat Crop by Spectral Reflectance Using ANN Algorithms. In P. Papajorgji & F. Pinet (Eds.), *Innovations and Trends in Environmental and Agricultural Informatics* (pp. 156–172). Hershey, PA: IGI Global. doi:10.4018/978-1-5225-5978-8.ch007

Karapistoli, E., Mampentzidou, I., & Economides, A. A. (2018). Environmental Monitoring Based on the Wireless Sensor Networking Technology: A Survey of Real-World Applications. In P. Papajorgji & F. Pinet (Eds.), *Innovations and Trends in Environmental and Agricultural Informatics* (pp. 196–251). Hershey, PA: IGI Global. doi:10.4018/978-1-5225-5978-8.ch009

Kasemsap, K. (2016). Multifaceted Applications of Green Supply Chain Management. In A. Jean-Vasile (Ed.), *Food Science, Production, and Engineering in Contemporary Economies* (pp. 327–354). Hershey, PA: IGI Global. doi:10.4018/978-1-5225-0341-5.ch014

Kateris, D., Gravalos, I., & Gialamas, T. (2019). Identification of Agricultural Crop Residues Using Non-Destructive Methods. In N. Razmjooy & V. Estrela (Eds.), *Applications of Image Processing and Soft Computing Systems in Agriculture* (pp. 114–144). Hershey, PA: IGI Global. doi:10.4018/978-1-5225-8027-0.ch005

Khalin, E. V., & Mikhaylova, E. E. (2019). Innovative Technologies for E-Learning of Safety of Agro-Industrial Production. In V. Kharchenko & P. Vasant (Eds.), *Advanced Agro-Engineering Technologies for Rural Business Development* (pp. 346–367). Hershey, PA: IGI Global. doi:10.4018/978-1-5225-7573-3.ch013

Kiberiti, B. S., Sanga, C. A., Mussa, M., Tumbo, S. D., Mlozi, M. R., & Haug, R. (2016). Farmers' Access and Use of Mobile Phones for Improving the Coverage of Agricultural Extension Service: A Case of Kilosa District, Tanzania. *International Journal of ICT Research in Africa and the Middle East*, *5*(1), 35–57. doi:10.4018/IJICTRAME.2016010103

Kitouni, I., Benmerzoug, D., & Lezzar, F. (2018). Smart Agricultural Enterprise System Based on Integration of Internet of Things and Agent Technology. *Journal of Organizational and End User Computing*, *30*(4), 64–82. doi:10.4018/JOEUC.2018100105

Komarchuk, D. S., Lysenko, V. P., Opryshko, O. O., & Pasichnyk, N. A. (2019). Monitoring the Condition of Mineral Nutrition of Crops Using UAV for Rational Use of Fertilizers. In V. Kharchenko & P. Vasant (Eds.), *Advanced Agro-Engineering Technologies for Rural Business Development* (pp. 293–319). Hershey, PA: IGI Global. doi:10.4018/978-1-5225-7573-3.ch011

Kothari, R., & Wani, K. A. (2019). Environmentally Friendly Slow Release Nano-Chemicals in Agriculture: A Synoptic Review. In R. Poonia, X. Gao, L. Raja, S. Sharma, & S. Vyas (Eds.), *Smart Farming Technologies for Sustainable Agricultural Development* (pp. 220–240). Hershey, PA: IGI Global. doi:10.4018/978-1-5225-5909-2.ch010

Koutsos, T., & Menexes, G. (2019). Economic, Agronomic, and Environmental Benefits From the Adoption of Precision Agriculture Technologies: A Systematic Review. *International Journal of Agricultural and Environmental Information Systems*, *10*(1), 40–56. doi:10.4018/IJAEIS.2019010103

Kozyrskyi, V., Zablodskiy, M., Savchenko, V., Sinyavsky, O., Yuldashev, R., Kalenska, S., & Podlaski, S. Z. (2019). The Magnetic Treatment of Water Solutions and Seeds of Agricultural Crops. In V. Kharchenko & P. Vasant (Eds.), *Advanced Agro-Engineering Technologies for Rural Business Development* (pp. 256–292). Hershey, PA: IGI Global. doi:10.4018/978-1-5225-7573-3.ch010

Krausp, V., & Ovik, G. B. (2019). Electric Robotized Organic Technology for Livestock Production on a Pasture Field. In V. Kharchenko & P. Vasant (Eds.), *Advanced Agro-Engineering Technologies for Rural Business Development* (pp. 180–198). Hershey, PA: IGI Global. doi:10.4018/978-1-5225-7573-3.ch007

Kumar, A. (2019). Characteristics of Various Soil Amendments: Soil Sustainability. In A. Rathoure (Ed.), *Amelioration Technology for Soil Sustainability* (pp. 1–12). Hershey, PA: IGI Global. doi:10.4018/978-1-5225-7940-3.ch001

Kumar, A. (2019). Impact Analysis of Amendment Application Under Diversified Agro-Ecological System: Sustainable Environment. In A. Rathoure (Ed.), *Amelioration Technology for Soil Sustainability* (pp. 135–150). Hershey, PA: IGI Global. doi:10.4018/978-1-5225-7940-3.ch008

Kumar, A. (2019). Red Mud (RM) and Soil Amelioration: Improvement in Soil Quality. In A. Rathoure (Ed.), *Amelioration Technology for Soil Sustainability* (pp. 151–167). Hershey, PA: IGI Global. doi:10.4018/978-1-5225-7940-3.ch009

Kumar, V. (2019). Synergism Between Microbes and Plants for Soil Contaminants Mitigation. In A. Rathoure (Ed.), *Amelioration Technology for Soil Sustainability* (pp. 101–134). Hershey, PA: IGI Global. doi:10.4018/978-1-5225-7940-3.ch007

Kuzman, B., Stegić, M., & Puškarić, A. (2016). Serbia and EFTA Contributions to Trade of Agroindustrial Products. In A. Jean-Vasile (Ed.), *Food Science, Production, and Engineering in Contemporary Economies* (pp. 355–364). Hershey, PA: IGI Global. doi:10.4018/978-1-5225-0341-5.ch015

Laguerre, O., & Chaomuang, N. (2019). Closed Refrigerated Display Cabinets: Is It Worth It for Food Quality? In P. Gaspar & P. da Silva (Eds.), *Novel Technologies and Systems for Food Preservation* (pp. 1–23). Hershey, PA: IGI Global. doi:10.4018/978-1-5225-7894-9.ch001

Lawton, S. (2016). Exploring the Meal Experience: Customer Perceptions of Dark-Dining. In A. Jean-Vasile (Ed.), *Food Science, Production, and Engineering in Contemporary Economies* (pp. 225–244). Hershey, PA: IGI Global. doi:10.4018/978-1-5225-0341-5.ch009

Lazăr, C., & Lazăr, M. (2016). Trends in the Evolution of Romania's Agricultural Resources in the Context of Sustainable Development. In A. Jean-Vasile (Ed.), *Food Science, Production, and Engineering in Contemporary Economies* (pp. 146–175). Hershey, PA: IGI Global. doi:10.4018/978-1-5225-0341-5.ch006

Lima de Aguiar, M., Gaspar, P. D., & da Silva, P. D. (2019). Frost Measuring and Prediction Systems for Demand Defrost Control. In P. Gaspar & P. da Silva (Eds.), *Novel Technologies and Systems for Food Preservation* (pp. 24–50). Hershey, PA: IGI Global. doi:10.4018/978-1-5225-7894-9.ch002

Marian, Z., & Balacescu, A. (2016). Food Consumption Patterns in Romanian Economy: A Framework. In A. Jean-Vasile (Ed.), *Food Science, Production, and Engineering in Contemporary Economies* (pp. 365–392). Hershey, PA: IGI Global. doi:10.4018/978-1-5225-0341-5.ch016

Monteiro, A. C., Iano, Y., França, R. P., & Razmjooy, N. (2019). WT-MO Algorithm: Automated Hematological Software Based on the Watershed Transform for Blood Cell Count. In N. Razmjooy & V. Estrela (Eds.), *Applications of Image Processing and Soft Computing Systems in Agriculture* (pp. 39–79). Hershey, PA: IGI Global. doi:10.4018/978-1-5225-8027-0.ch002

Naraine, L. (2019). *Optimizing the Use of Farm Waste and Non-Farm Waste to Increase Productivity and Food Security: Emerging Research and Opportunities* (pp. 1–223). Hershey, PA: IGI Global. doi:10.4018/978-1-5225-7934-2

Parece, T. E., & Campbell, J. B. (2017). Geospatial Evaluation for Urban Agriculture Land Inventory: Roanoke, Virginia USA. *International Journal of Applied Geospatial Research*, 8(1), 43–63. doi:10.4018/IJAGR.2017010103

Pathare, P. B., Roskilly, A. P., & Jagtap, S. (2019). Energy Efficiency in Meat Processing. In P. Gaspar & P. da Silva (Eds.), *Novel Technologies and Systems for Food Preservation* (pp. 78–107). Hershey, PA: IGI Global. doi:10.4018/978-1-5225-7894-9.ch004

Peças, P., Fonseca, G. M., Ribeiro, I. I., & Sørensen, C. G. (2019). Automation of Marginal Grass Harvesting: Operational, Economic, and Environmental Analysis. In R. Poonia, X. Gao, L. Raja, S. Sharma, & S. Vyas (Eds.), *Smart Farming Technologies for Sustainable Agricultural Development* (pp. 106–146). Hershey, PA: IGI Global. doi:10.4018/978-1-5225-5909-2.ch006

Plazas, J. E., Bimonte, S., De Sousa, G., & Corrales, J. C. (2019). Data-Centric UML Profile for Wireless Sensors: Application to Smart Farming. *International Journal of Agricultural and Environmental Information Systems*, *10*(2), 21–48. doi:10.4018/IJAEIS.2019040102

Popović, V. Ž., Subić, J. V., & Kljajić, N. Ž. (2016). The Role of Irrigation in the Development of Agriculture: Srem District (Serbia). In A. Jean-Vasile (Ed.), *Food Science, Production, and Engineering in Contemporary Economies* (pp. 102–124). Hershey, PA: IGI Global. doi:10.4018/978-1-5225-0341-5.ch004

Raja, L., & Vyas, S. (2019). The Study of Technological Development in the Field of Smart Farming. In R. Poonia, X. Gao, L. Raja, S. Sharma, & S. Vyas (Eds.), *Smart Farming Technologies for Sustainable Agricultural Development* (pp. 1–24). Hershey, PA: IGI Global. doi:10.4018/978-1-5225-5909-2.ch001

Rajinikanth, V., Arunmozhi, S., Raja, N. S., Varthini, B. P., & Thanaraj, K. P. (2019). Examination of Plant/Weed Image Dataset Using a Hybrid Image Processing Tool. In N. Razmjooy & V. Estrela (Eds.), *Applications of Image Processing and Soft Computing Systems in Agriculture* (pp. 159–183). Hershey, PA: IGI Global. doi:10.4018/978-1-5225-8027-0.ch007

Ramar, K., & Gurunathan, G. (2017). Semantic Web Based Agricultural Information Integration. *International Journal of Agricultural and Environmental Information Systems*, *8*(3), 39–51. doi:10.4018/IJAEIS.2017070103

Ramar, K., & Gurunathan, G. (2018). Semantic Web-Based Agricultural Information Integration. In P. Papajorgji & F. Pinet (Eds.), *Innovations and Trends in Environmental and Agricultural Informatics* (pp. 86–104). Hershey, PA: IGI Global. doi:10.4018/978-1-5225-5978-8.ch004

Rathoure, A. K. (2019). Soil Quality and Soil Sustainability: Sustainable Agro-Ecosystem Management. In A. Rathoure (Ed.), *Amelioration Technology for Soil Sustainability* (pp. 46–58). Hershey, PA: IGI Global. doi:10.4018/978-1-5225-7940-3.ch004

Rathoure, A. K. (2019). Soil Quality of Tripura State of India: Sustainable Environment. In A. Rathoure (Ed.), *Amelioration Technology for Soil Sustainability* (pp. 195–204). Hershey, PA: IGI Global. doi:10.4018/978-1-5225-7940-3.ch012

Rathoure, A. K. (2019). Soil Sampling, Analysis, and Rock Phosphate Amendments: Good Practices for Soil Sustainability. In A. Rathoure (Ed.), *Amelioration Technology for Soil Sustainability* (pp. 32–45). Hershey, PA: IGI Global. doi:10.4018/978-1-5225-7940-3.ch003

Rathoure, K. P. (2019). Amelioration Technology for Agricultural Efficiency: Biochar and Compost Amendments for Soil sustainability. In A. Rathoure (Ed.), *Amelioration Technology for Soil Sustainability* (pp. 13–31). Hershey, PA: IGI Global. doi:10.4018/978-1-5225-7940-3.ch002

Rathoure, K. P. (2019). Impacts of Phosphate Amendments at a Contaminated Site: Soil Sustainability. In A. Rathoure (Ed.), *Amelioration Technology for Soil Sustainability* (pp. 90–100). Hershey, PA: IGI Global. doi:10.4018/978-1-5225-7940-3.ch006

Rathoure, K. P. (2019). Soil Quality Near Indian Oil Corporation Limited Pol Depot Ahmednagar, Maharashtra, State of India. In A. Rathoure (Ed.), *Amelioration Technology for Soil Sustainability* (pp. 168–178). Hershey, PA: IGI Global. doi:10.4018/978-1-5225-7940-3.ch010

Razmjooy, N., Estrela, V. V., & Loschi, H. J. (2019). A Survey of Potatoes Image Segmentation Based on Machine Vision. In N. Razmjooy & V. Estrela (Eds.), *Applications of Image Processing and Soft Computing Systems in Agriculture* (pp. 1–38). Hershey, PA: IGI Global. doi:10.4018/978-1-5225-8027-0.ch001

Rodrigues, G., & Guimarães, A. M. (2018). Could NoSQL Replace Relational Databases in FMIS? In P. Papajorgji & F. Pinet (Eds.), *Innovations and Trends in Environmental and Agricultural Informatics* (pp. 38–57). Hershey, PA: IGI Global. doi:10.4018/978-1-5225-5978-8.ch002

Seetharaman, K. (2019). Applications of Image Compression on Agricultural Image Data Analysis. In N. Razmjooy & V. Estrela (Eds.), *Applications of Image Processing and Soft Computing Systems in Agriculture* (pp. 208–241). Hershey, PA: IGI Global. doi:10.4018/978-1-5225-8027-0.ch009

Senkevich, S., Ivanov, P. A., Lavrukhin, P. V., & Yuldashev, Z. (2019). Theoretical Prerequisites for Subsurface Broadcast Seeding of Grain Crops in the Conditions of Pneumatic Seed Transportation to the Coulters. In V. Kharchenko & P. Vasant (Eds.), *Advanced Agro-Engineering Technologies for Rural Business Development* (pp. 28–64). Hershey, PA: IGI Global. doi:10.4018/978-1-5225-7573-3.ch002

Senkevich, S. E., Sergeev, N. V., Vasilev, E. K., Godzhaev, Z. A., & Babayev, V. (2019). Use of an Elastic-Damping Mechanism in the Tractor Transmission of a Small Class of Traction (14 kN): Theoretical and Experimental Substantiation. In V. Kharchenko & P. Vasant (Eds.), *Advanced Agro-Engineering Technologies for Rural Business Development* (pp. 149–179). Hershey, PA: IGI Global. doi:10.4018/978-1-5225-7573-3.ch006

Sima, V., & Gheorghe, I. G. (2016). Perception of Romanian Consumers on Ecological Products. In A. Jean-Vasile (Ed.), *Food Science, Production, and Engineering in Contemporary Economies* (pp. 203–224). Hershey, PA: IGI Global. doi:10.4018/978-1-5225-0341-5.ch008

Simonovic, Z., & Vukovic, P. (2016). Characteristics Development of Agriculture and Agricultural Policy Southeast European Countries. In A. Jean-Vasile (Ed.), *Food Science, Production, and Engineering in Contemporary Economies* (pp. 275–293). Hershey, PA: IGI Global. doi:10.4018/978-1-5225-0341-5.ch011

Singh, D., & Sharma, D. (2019). Prognosis for Crop Yield Production by Data Mining Techniques in Agriculture. In N. Razmjooy & V. Estrela (Eds.), *Applications of Image Processing and Soft Computing Systems in Agriculture* (pp. 145–158). Hershey, PA: IGI Global. doi:10.4018/978-1-5225-8027-0.ch006

Singh, T., & Vinod, D. N. (2019). Intelligent Farming With Surveillance Agribot. In N. Razmjooy & V. Estrela (Eds.), *Applications of Image Processing and Soft Computing Systems in Agriculture* (pp. 272–296). Hershey, PA: IGI Global. doi:10.4018/978-1-5225-8027-0.ch011

Stancu, A. (2016). Correlations and Patterns of Food and Health Consumer Expenditure. In A. Jean-Vasile (Ed.), *Food Science, Production, and Engineering in Contemporary Economies* (pp. 44–101). Hershey, PA: IGI Global. doi:10.4018/978-1-5225-0341-5.ch003

Stefanovskiy, V. (2019). Processes and Technological Systems for Thawing of Fish. In P. Gaspar & P. da Silva (Eds.), *Novel Technologies and Systems for Food Preservation* (pp. 51–77). Hershey, PA: IGI Global. doi:10.4018/978-1-5225-7894-9.ch003

Tagliolato, P., & Manfredini, F. (2018). Discovering Regularity Patterns of Mobility Practices Through Mobile Phone Data. In P. Papajorgji & F. Pinet (Eds.), *Innovations and Trends in Environmental and Agricultural Informatics* (pp. 173–195). Hershey, PA: IGI Global. doi:10.4018/978-1-5225-5978-8. ch008

Tikhomirov, D., Vasiliev, A., & Dudin, S. (2019). Energy-Saving Electrical Installations for Heat Supply of Agricultural Objects. In V. Kharchenko & P. Vasant (Eds.), *Advanced Agro-Engineering Technologies for Rural Business Development* (pp. 96–122). Hershey, PA: IGI Global. doi:10.4018/978-1-5225-7573-3.ch004

Umamaheswari, S. (2019). Internet of Things Practices for Smart Agriculture. In R. Poonia, X. Gao, L. Raja, S. Sharma, & S. Vyas (Eds.), *Smart Farming Technologies for Sustainable Agricultural Development* (pp. 67–92). Hershey, PA: IGI Global. doi:10.4018/978-1-5225-5909-2.ch004

Vasiliev, A. N., Vasiliev, A. A., Budnikov, D., & Samarin, G. (2019). Analysis of Experimental Research Results Obtained for Grain Drying With Electrically Activated Air. In V. Kharchenko & P. Vasant (Eds.), *Advanced Agro-Engineering Technologies for Rural Business Development* (pp. 230–255). Hershey, PA: IGI Global. doi:10.4018/978-1-5225-7573-3.ch009

Vasiliev, A. N., Vasiliev, A. A., Budnikov, D., Tikhomirov, D., & Ospanov, A. B. (2019). Control and Optimized Management of Grain Drying in Forced-Aerated Bins. In V. Kharchenko & P. Vasant (Eds.), *Advanced Agro-Engineering Technologies for Rural Business Development* (pp. 199–229). Hershey, PA: IGI Global. doi:10.4018/978-1-5225-7573-3.ch008

Verma, S., Chug, A., Singh, A. P., Sharma, S., & Rajvanshi, P. (2019). Deep Learning-Based Mobile Application for Plant Disease Diagnosis: A Proof of Concept With a Case Study on Tomato Plant. In N. Razmjooy & V. Estrela (Eds.), *Applications of Image Processing and Soft Computing Systems in Agriculture* (pp. 242–271). Hershey, PA: IGI Global. doi:10.4018/978-1-5225-8027-0.ch010

Verma, V. K., & Jain, T. (2019). Soft-Computing-Based Approaches for Plant Leaf Disease Detection: Machine-Learning-Based Study. In N. Razmjooy & V. Estrela (Eds.), *Applications of Image Processing and Soft Computing Systems in Agriculture* (pp. 100–113). Hershey, PA: IGI Global. doi:10.4018/978-1-5225-8027-0.ch004

Vieito, C., Pires, P., & Velho, M. V. (2019). Pinus Pinaster Bark Composition and Applications: A Review. In P. Gaspar & P. da Silva (Eds.), *Novel Technologies and Systems for Food Preservation* (pp. 174–189). Hershey, PA: IGI Global. doi:10.4018/978-1-5225-7894-9.ch008

Voland, P., & Asche, H. (2018). Processing and Visualizing Floating Car Data for Human-Centered Traffic and Environment Applications: A Transdisciplinary Approach. In P. Papajorgji & F. Pinet (Eds.), *Innovations and Trends in Environmental and Agricultural Informatics* (pp. 105–128). Hershey, PA: IGI Global. doi:10.4018/978-1-5225-5978-8.ch005

Wutofeh, W. C. (2019). ICTs and Improvement of Agriculture in the North West Region of Cameroon. *International Journal of Information Communication Technologies and Human Development*, *11*(1), 1–19. doi:10.4018/IJICTHD.2019010101

Yadav, V. K., & Pandita, P. R. (2019). Fly Ash Properties and Their Applications as a Soil Ameliorant. In A. Rathoure (Ed.), *Amelioration Technology for Soil Sustainability* (pp. 59–89). Hershey, PA: IGI Global. doi:10.4018/978-1-5225-7940-3.ch005

Yıkmış, S. (2019). Uses of Non-Thermal Treatment Technologies in Liquid Foodstuff. In P. Gaspar & P. da Silva (Eds.), *Novel Technologies and Systems for Food Preservation* (pp. 160–173). Hershey, PA: IGI Global. doi:10.4018/978-1-5225-7894-9.ch007

Yudaev, I. V., Charova, D., Feklistov, A., Mashkov, S., Kryuchin, P., Vasilyev, S., ... Armenyanov, N. (2019). Research of Green Vegetable Cultivation Technology Under Photoculture Conditions in Irradiation Chamber. In V. Kharchenko & P. Vasant (Eds.), *Advanced Agro-Engineering Technologies for Rural Business Development* (pp. 368–395). Hershey, PA: IGI Global. doi:10.4018/978-1-5225-7573-3.ch014

Yudaev, I. V., Daus, Y., Kokurin, R., Prokofyev, P. V., Gamaga, V., & Armenyanov, N. (2019). Electro-Impulse Irreversible Plant Tissue Damage as Highly Efficient Agricultural Technology. In V. Kharchenko & P. Vasant (Eds.), *Advanced Agro-Engineering Technologies for Rural Business Development* (pp. 396–430). Hershey, PA: IGI Global. doi:10.4018/978-1-5225-7573-3.ch015

Yuferev, L., Sokolov, A., & Mironyuk, S. S. (2019). UV-Based Indoor Disinfecting System. In V. Kharchenko & P. Vasant (Eds.), *Advanced Agro-Engineering Technologies for Rural Business Development* (pp. 65–95). Hershey, PA: IGI Global. doi:10.4018/978-1-5225-7573-3.ch003

Zaharia, M., & Gogonea, R. (2016). Food Consumption Expenditure and Standard of Living in Romania. In A. Jean-Vasile (Ed.), *Food Science, Production, and Engineering in Contemporary Economies* (pp. 245–274). Hershey, PA: IGI Global. doi:10.4018/978-1-5225-0341-5.ch010

Zubovic, J., & Pavlovic, D. M. (2016). Youth Employability in WB Countries: Can They Look Up to Youth in Developed Countries? In A. Jean-Vasile (Ed.), *Food Science, Production, and Engineering in Contemporary Economies* (pp. 315–326). Hershey, PA: IGI Global. doi:10.4018/978-1-5225-0341-5.ch013

About the Contributors

Proshikshya Mukherjee is currently pursuing PhD in KIIT Deemed to be University. M. Tech (Communication System Engineering) is working as faculty at the School of Electronics Engineering, KIIT University, her field of interest includes Cloud Computing, Wireless Sensor Network and IoT. Proshikshya has published numbers of Research papers in peer reviewed international journals and conferences.

Prasant Kumar Pattnaik, Ph.D. (Computer Science), Fellow IETE, Senior Member IEEE, is Professor at the School of Computer Engineering, KIIT Deemed to be University, Bhubaneswar. He has more than a decade of teaching research experience. Dr. Pattnaik has published numbers of Research papers in peer reviewed international journals and conferences. His researches areas are Cloud Computing, Mobile Computing and Brain Computer Interface. He authored many computer science books in field of Robotics, Turing Machine, Cloud Computing, and Mobile Computing.

Surya Narayan Panda, Director Research, Chitkara University, Punjab, India, is Ph.D. Computer Science, working towards development of innovative technologies and product based on Internet of things and Cloud Computing. He is expertized in Cyber security, Networking, Advanced Computer Network, Machine Learning and Artificial Intelligence. He has filed 8 patents, 55 international publications in the relevant area and involved in Internet of things healthcare devices like Portable Intensive Care Unit, Digital Laryngoscope etc. He has developed the prototype of Smart Portable Intensive Care Unit through which doctor can provide the immediate virtual medical assistance to emergency cases in ambulance and won prestigious Millennium Alliance Award from FICCI in 2017 and seed funding for his project. He is also working on a project "Cyber Technology Communication for Women

Safety" which is funded by Ministry of Science and Technology, Govt. Of India and another project "Remote Vital Information and Surveillance System for Elderlyand Disabled Persons" which is again funded by Ministry of Science and Technology, Govt. of India.

* * *

P. Alli received the B.E. degree in Electronics & Communication and Engineering from Madras University and M.S. degree in Birla Institute of Technology and Science and Ph.D. degree in Image Processing in Madurai Kamaraj University, Madurai, Tamil Nadu, India. She is the Professor and Head of the Department of Computer Science and Engineering in Velammal College of Engineering & Technology. She has authored more than 35 publications in journals and 20 publications in conferences. She is the Principal Investigator of the funded project from DST, India titled "Efficient e-waste Management through Coordinated Web Service using WS-Dependable Space (DST No: DST/SSTP/TN/194/2011)". She is the Principal Investigator of the funded projects from MNRE, titled "CST System for Process Heat / Heat Applications" and also received funds from various agencies like DST–IEDC and IE(I). Her research interests include Image Processing, Networking, and Information Security.

M. Vasim Babu received his B.E. degree in the discipline of Electronics and Communication Engineering from Sethu Institute of Technology, India, M.E Degree in the discipline of Communication Systems from K.L.N college of engineering, India, the Ph.D. degree in the discipline of Information and Communication Engineering from Anna University, India. His area of interest is in Wireless Sensor Networks, Adhoc network, ANFIS and Signal processing. He is an active member of International Association of Engineers, the Seventh sense research group, International Association of Computer Science and Information Technology, Universal Association of Computer and Electronics Engineers and the Institute of engineers. Currently, he is working as an associate professor, Department of ECE in KKR and KSR Institute of Technology and Sciences, Guntur, AP.

G. Vinoth Chakkaravarthy received the B.E. degree in Computer Science and Engineering from Madurai Kamaraj University and M.E. degree in Computer Science and Engineering in Anna University, India. and Ph.D.

degree in Network Security from Anna University. He is an Associate Professor of Computer Science and Engineering in Velammal College of Engineering & Technology. He has authored more than 12 publications in journals and conferences. He is the Co-PI of the funded project from DST, India, titled "Efficient e-waste Management through Coordinated Web Service using WS-Dependable Space (DST No: DST/SSTP/TN/194/2011)". His research interests include Cryptography, Network Security, data security and privacy, IoT and VANET security.

Subba Reddy Chavva completed B.Tech in Information Technology from MITS, Madanapalle in 2013 and M.Tech in Software Engineering from JNTUACEA, Anantapuramu in 2016. Currently, pursuing the Ph.D. degree in Computer Science at Vellore Institute of Technology-AP University, Amaravati. His research interests include wireless Sensor Networks and Soft Computing.

Ramgopal Kashyap's areas of interest are image processing, pattern recognition, and machine learning. He has published many research papers, and book chapters in international journals and conferences like Springer, Inderscience, Elsevier, ACM, and IGI-Global indexed by Science Citation Index (SCI) and Scopus (Elsevier). He has Reviewed Research Papers in the Science Citation Index Expanded, Springer Journals and Editorial Board Member and conferences programme committee member of the IEEE, Springer international conferences and journals held in countries: Czech Republic, Switzerland, UAE, Australia, Hungary, Poland, Taiwan, Denmark, India, USA, UK, Austria, and Turkey. He has written many book chapters published by IGI Global, USA, Elsevier and Springer.

Raja Lavanya received the B.E. degree and M.E. degree in Computer science and Engineering in Anna University, Chennai, Tamil Nadu, India. She is an Assistant Professor of Computer Science and Engineering in Thiagarajar college of Engineering. She has authored more than 5 publications. Her research interests include Cryptography, Network Security, data security and privacy and IoT security.

Shaswati Patra is an assistant professor in School of Computer Engineering at Kalinga Institute Of Industrial Technology Completed M.Tech from NIT Durgapur, Completed B.Tech under BPUT.

Sarita Tripathy received B.E degree from National Institute of Science and Technology, Berhampur Odisha, in year 2003 and M.Tech degree from College Engineering and Technology Bhubaneswar in Computer Science and Engineering in the year 2008. She is currently pursuing her Ph.D. in the field of Data Mining from KIIT University Bhubaneswar Odisha. She has more than ten years of teaching experience, moreover, she is presently working as Assistant Professor in KIIT University Bhubaneswar. Her research area includes: Data Mining, Data Analytics and Soft computing.

Nagesh Mallaiah Vaggu is presently working as Assistant Professor in the Department of Computer Science and Engineering, ASHOKA Institute of Engineering and Technology. Currently, pursuing my PhD in part-time mode from VIT AP University. Area of Research is Wireless Sensor Networks.

Index

A

actuator network 78
agricultural field 4-6
Agriculture 1-4, 9, 41-42, 44, 48, 86-88,
    91-92, 94, 96-97, 101, 117
alert message 3-5
Antithetic Sampling 52, 56-57, 65
Application Layer 2, 4-5, 32

B

Body Sensor Network 24
Business 19, 21, 102, 108-109

C

CC2420 Radio Chip 52, 56
Clustering 63, 77-78
clusters 79-81
communication 2, 4-5, 9, 12-13, 23-24, 27,
    29, 32-33, 42, 48, 57, 63, 65, 77-78,
    80, 91, 102
consumption 4, 42-43, 45, 47-48, 53-57,
    62-64, 73, 100
crops 1-2, 87, 90, 95, 99

D

Data Dissemination 46-47
Data Mining 87, 91, 94
Decision Support System 86
Dijkstra's Algorithm 78, 80-81, 83
Discrete Radio Model 65
Distillation 107-108, 115-117

E

E-Health 8, 10, 14, 22-23, 25
ELECTRE III 99-102, 104, 106, 109, 118

F

farmer 3-6, 101

G

GIS 25, 86

H

high power 53

I

IHH 40
Image processing 87, 91, 96
Internet of Everything 14, 16, 18-20
Internet of Things 1-4, 8, 10-11, 35, 40,
    87, 91
Intra-Cluster Communication 77-78, 80
IoT 1-6, 8, 10, 12, 22-23, 25-26, 32-34, 36,
    40, 86, 91, 97

L

LEACH 77-81, 83
LEACH-V 77-78, 80, 82-83
LEACH-VD 77, 80, 82-84
Localization 11, 36, 52-57, 59, 61-62,
    66-71, 73

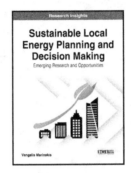

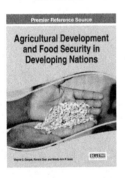
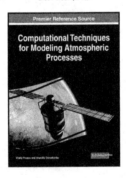
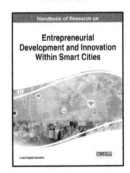

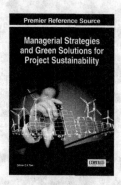